4th International Conference on Sustainable Practices and Innovations in Civil Engineering (SPICE 2023)

AA003630

MATEC Web of Conferences Volume 384 (2023)

Chennai, India
23 - 24 March 2023

Editor:

S. Natarajan

ISBN: 978-1-7138-8338-8

Printed from e-media with permission by:

Curran Associates, Inc.
57 Morehouse Lane
Red Hook, NY 12571

Some format issues inherent in the e-media version may also appear in this print version.

This work is licensed under a Creative Commons Attribution 4.0 International License. License details: http://creativecommons.org/licenses/by/4.0/.

No changes have been made to the content of these proceedings. There may be changes to pagination and minor adjustments for aesthetics.

Printed with permission by Curran Associates, Inc. (2024)

For additional information, please contact EDP Sciences – Web of Conferences
at the address below.

EDP Sciences – Web of Conferences
17, Avenue du Hoggar
Parc d'Activité de Courtabœuf
BP 112
F-91944 Les Ulis Cedex A
France

Phone: +33 (0) 1 69 18 75 75
Fax: +33 (0) 1 69 28 84 91

contact-edps@webofconferences.org

Additional copies of this publication are available from:

Curran Associates, Inc.
57 Morehouse Lane
Red Hook, NY 12571 USA
Phone: 845-758-0400
Fax: 845-758-2633
Email: curran@proceedings.com
Web: www.proceedings.com

TABLE OF CONTENTS

edp sciences

MATEC Web of Conferences

Statement of Peer review

In submitting conference proceedings to MATEC Web of Conferences, I certify to the Publisher that I adhere to the **Policy on Publishing Integrity** of the journal in order to safeguard good scientific practice in publishing.

1. All articles have been subjected to peer review administered by the proceedings editors.
2. Reviews have been conducted by expert referees, who have been requested to provide unbiased and constructive comments aimed, whenever possible, at improving the work.
3. Proceedings editors have taken all reasonable steps to ensure the quality of the materials they publish and their decision to accept or reject a paper for publication has been based only on the merits of the work and the relevance to the journal.

Title, date and place of the conference Sustainable Practices and Innovations In Civil ENGineering (SPICE 2023) on 23rd and 24th March 2023 Sri Sivasubramaniya Nadar College of ENGineering (SSN CE) Chennai, Kalavakkam,-603110, Chengalpattu Dist, TamilNadu, India.

Proceedings editor(s):

Date and editor's signature 20/10/2023.

Dept of Civil Engineering
SSN College of Engineering
Kalavakkam - 603 110

DEPARTMENT OF CIVIL ENGINEERING ★ SSNCE KALAVAKKAM

MATEC Web of Conferences **384**, 01001 (2023) https://doi.org/10.1051/matecconf/202338401001
SPICE 2023

A study on reduction of air pollution by adding mesoporous adsorbent in road pavements

P. Indhiradevi[1], *B.S.Meenakshi*[2], *M. Ram Vivekanandhan*[3], *G. Keerthigha* [4],*Nekkil Chettri*[5],*Poorna pushkalapriya*[6]

[1,2&3]Assistant Professor, Department of Civil Engineering, KPR Institute of Engineering and Technology, Coimbatore. Email id: Indhiradevi.p@kpriet.ac.in
[4,5&6]Under Graduate Student, Department of Civil Engineering, KPR Institute of Engineering and Technology, Coimbatore.

Abstract. The amount of dangerous pollutants such carbon dioxide, Sulphur dioxide, nitrogen dioxide, hydrocarbons, and fine particulate matter increases as a result of vehicle pollution. Vehicle pollution, which is caused by harmful substances released into the environment by motor vehicles, helps to create air pollution. While the impacts of air pollution on objects, plants, and animals can be quantified, the effects on human health can only be inferred from epidemiological data. Additionally, the observations of the impacts of air pollution are complicated by the health effects of smoking and other lifestyle factors and exposures. According to all available data, air pollutants pose such a risk to people's health and welfare that their control is required. The addition of the mesoporous adsorbent effectively consumes the air pollutants and reduces air pollution and limits the vehicular emission control timely. When the mesoporous adsorbent is added along with the materials used in road pavement, no matter how many vehicles move on the road this material simultaneously adsorbs the vehicular emissions. This study results in major reductions in vehicle pollutant emissions with net saving in life and substantial improvement in the environmental conditions.

1 Introduction

The contamination of the atmosphere by wastes or byproducts that are gaseous, liquid, or solid can be harmful to human health as well as the welfare of plants and animals, attack building materials, impair vision, or release objectionable odours. The bulk of pollutants originate from man-made sources, but some also come from hot springs, coniferous forests, and volcanoes. These have a very minor impact on pollution compared to emissions from industry, the use of power and heat, combustion engines, and waste disposal. Vehicle emissions are responsible for around 60% of air pollution. The main source of air pollution emissions that result from human activity is fuel combustion.
Carbon monoxide (CO), carbon dioxide (CO2), nitrogen oxide (NOx) in the forms of nitric oxide (NO) and nitrogen dioxide (NO2), hydrocarbons (HC), sulphur oxides (SOX) in the forms of sulphur dioxide (SO2) and sulphur trioxide (SO3), and particulate matter are the main pollutants released from motor vehicles (PM10). According to a survey, the nation will

© The Authors, published by EDP Sciences. This is an open access article distributed under the terms of the Creative Commons Attribution License 4.0 (https://creativecommons.org/licenses/by/4.0/).

spew almost 68 million tonnes of pollution into the sky by the year 2020. And as a result of these emissions, ozone and particle production, acid deposition, and reduced vision are all brought about. All of these have an impact on both the environment and human health.

In order to prevent this pollution, we can relate all the issues and use materials like mesoporous adsorbents. In this project, we'll conduct research on the usage of mesoporous adsorbent in road building to reduce air pollution, which is primarily brought on by vehicle emissions

2 Mesoporous adsorbent

The substance is created by combining sodium bicarbonate's gel-forming characteristics with regulated porosity and alginate's low temperature breakdown.

2.1 Properties

- Less expensive and more environmentally friendly to create.
- Porous surface catches and absorbs particulate matter from vehicle exhaust fumes;
- Due to its colloidal characteristic, the substance creates an insoluble gel when divalent metals are present.
- By changing the pore volume appropriately, flexibility is increased.
- Sustainable and environmentally friendly materials, high-efficiency materials, and materials that don't need stability by thermal treatment at high temperatures
- The pore volume can be adjusted accordingly increasing its flexibility.
- Highly efficient material.
- don't need stability by thermal treatment at high temperatures
- Has high versatility
- Material exhibits good photo degradation performance.

2.2 Applications

- Development of low cost and effective adsorbents.
- Adsorbing particulate matter from the pollutants.
- Converts potentially toxic waste into safe materials.
- Pollution can be managed at low costs.
- Can be used for mechanical and chemical stability.

2.3 Objective and methodology of the study

The objective of this Project is,

a) To enhance substantial improvement in the environmental conditions.
b) To prevent adverse conditions by the vehicular emissions exposed to the atmosphere.
c) To examine and make advantage of the new material's particle-capturing abilities.
d) To offer for the mitigation, control, and prevention of air pollution.
e) To use the adsorbent and improve the efficiency of air pollution control.

Methodology of the Project

a) To find out the mixing design for laying road.
b) To find out the procedure for mixing mesoporous adsorbent
c) To identify and analyze the changes in the mixture.
d) To measure the amount of particulate matter captured by that adsorbent.

e) To observe the effectiveness of the adsorbent used in road pavement..

3 Literature study

By combining the gel properties of alginate and the decomposition of sodium bicarbonate, Alessandra Zanoletti, Ivano Vassura, Elisa Venturini, and Matteo Monai created a mesoporous adsorbent material for their study on a new porous hybrid material derived from silica fume and alginate for sustainable pollutants reduction. The product obtained exhibited the properties such as versatile, amorphous, organic inorganic hybrid, high performance as an adsorbent and highly sustainable. When the material was placed near the engine exhaustion it was observed that almost 94% of the oxides emitted from fuel combustion were captured. These captured pollutants can be seen on the material formed as a layer. So, it was concluded that the material is highly efficient than other adsorbents.

In order to treat sodium alginate films with CaCl2, Jong Whan Rhim used two different techniques: he added CaCl2 directly to the film-making solution (when mixing the films) and he submerged the alginate films in the CaCl2 solutions (immersion films). According to the observations, the alginate films showed permeability, a uniform, translucent appearance, colloidal characteristic, and were non-toxic, biodegradable, and compatible with living things. These films are frequently utilised in the pharmaceutical industry for controlled medication release, wound healing, and drug delivery.

Mesoporous adsorbent for CO2 capture application under mild conditions is a research endeavour by two authors, including Muhammad Arif Ab Aziz: A summary Carbon dioxide (CO2) capture and storage have been recommended as a successful technique to lower global greenhouse gas emissions, according to an article in the Journal of Environmental Chemical Engineering. The mesoporous adsorbent generated from biomass sources is utilised to capture the CO2 at low adsorbent temperature and ambient pressure. It showed quick CO2 adsorption kinetics and steady cycling performance. The adsorbent substance is seen to have great promise and to be working perfectly as a capturing substance. High surface area, low energy consumption, low cost, and wide availability are desirable characteristics of such adsorbents.

Understanding Air Pollution from Induced Traffic During and After the Construction of a New Highway: Case Study of Highway 25 in Montreal by Md. Shohel Reza Amin, Umma Tamima, and Luis Amador Jimenez inferred that there is an increase in the concentration of contaminants with the increase in traffic density. Four scenarios of peak and off peak hour traffic and traffic density were considered during and after the construction of project. It was observed that there was less traffic in the neighborhood areas and as a result there is a decrease in the concentration of nitrogen oxides. There was an increase in the traffic density within the close vicinity leading to increase in the concentration of NO2 levels during and after the construction.

From their work on Reducing Urban Runoff Pollution Using Porous Concrete Containing Mineral Adsorbents, Ehsan Teymouri, Hojat Karami, Sayed Farhad Mousavi, and Saeed Farzin explained that adding minerals to road pavements promotes the adsorption of pollutants. Porous concrete has been employed in the sidewalks and surface pavements of this study. These have varying levels of adsorption capacity for pollutants (0, 5, 10, and 15%). Compressive strength, permeability, porosity, and the capacity to enhance runoff quality are the mechanical attributes. According to the findings, adding fine grains increased compressive strength and improved runoff quality, but decreased permeability and porosity.

All adsorbents improved runoff quality, and the improvement persisted at greater added percentages. However, the reduction of EC, TDS, and NaCl concentrations was not significantly impacted by porous concrete. The samples that contained zeolite and LECA produced the greatest outcomes, according to evaluation. The L15-0, Z15-10, and L15-20 treatments (containing 0, 10, and 20% fine-grains and 15% adsorbent) improved the TSS (75.8, 79.1, and 84.6%), COD (87.1, 82.6, and 89.3%), and BOD (88.1, 87.3, and 90.7%) and had the highest pollution reduction.

Anupriya B. and Sriman Narayanan K. looked into their research on smog-absorbing pavement. They have investigated this with the primary goals of preserving the ecosystem and reducing air pollution in heavily contaminated areas. Two photo catalytic additives were combined, coated, and sprayed over the pavement to create smog-absorbing pavement. These advantageous traits for the environment were seen. After utilising titanium dioxide to build the paver blocks, they were stored for one to two weeks. Later, each pavement underwent a scanning electron microscopic examination. The findings suggest that the pavements contain various amounts of pollutants, including carbon dioxide and nitrogen oxides. A certain quantity of air contaminants from car emissions can be absorbed by Tio2.

In their study Road Materials and Pavement Design, Taylor and Francis Ltd. made contributions on mix design, soil stabilisation, and environmental aspects of handling and reusing road materials, as well as the mechanical, thermal, chemical, and physical properties of mixes, asphalt concrete, cement concrete, unbound granular materials, bitumen's, additives, bituminous mixes, soils, and new and innovative materials. Research was done on pavement performance based on structural examination, stress, strain, and thermal characterization and computation, vehicle/road interaction, climatic effects, numerical and analytical modelling. Consideration is given to the soil as well as the various levels of the road.

In their paper Performance of Recycled Asphalt Pavement in Gravel Roads, Scott Koch, Khaled Ksaibati, and George Huntington examined the usage of additional recycled asphalt pavement (RAP) in roads. Recycled asphalt was added to the pavements to test them. They were built and watched over in two counties. A dustometer was used to measure the loss of dust. A method created by the U.S. Army Corps of Engineers in Unsurfaced Road Maintenance Management was utilised to conduct the test and assess its success. After the pavement is installed, its performance is checked, and statistical data was gathered. Data was gathered on how much less dust is generated when cars drive on pavement. The recycled asphalt pavement and the control sections made of gravel were contrasted. Conclusions showed that recycled asphalt pavements are more efficient at reducing dust without compromising the road's usability. Fundamental inferences and suggestions about RAP's capacity to reduce dust were made possible by this comparison. Additionally, it was discovered throughout the experiment that adding recycled asphalt to pavements helps save natural resources by reducing waste from petroleum-based products. The usage of it on unpaved roads, cycling paths, parking lots, residential driveways, back fills, and pipe beds were also created as further ideas.

In their work on Fly ash exploited in pavement layers in ecologically acceptable ways, Athanasopoulou and George Kollaros noted that environmentally friendly items must be used to lessen the worry for ozone depletion, solid and liquid waste disposal, and contaminants. Fly ash, a waste product, is used for the study in place of more expensive clays. Depending on the kind of soil, it was applied to the pavements at quantities between 8% and 12%. To enhance the engineering qualities of soil, the concentrations were appropriately changed. One of the most critical aspects of construction is soil stability. Fly ash can effectively stabilise the soil by being added. As a result, soil's shear strength and shrinkage characteristics both improve. It is possible to attain some desired features, like increased resistance, decreased flexibility, decreased permeability, decreased pavement thickness, and enhanced compaction. The use of fly ash increased the soil strength and unconfined compressive strength, according to the experimental data. The conclusion reached was that fly ash is an economical and effective building material.

3.1 Mesoporous adsorbent in road pavement

The first step carried out is planning of the road pavement. It is evaluated by its usage and needs in that particular area. The second step followed is the design. Alignment, shape and direction of road are pointed by the surveyors after they physically visit the site. Next the earthworks are carried out to provide a firm and stable foundation. Drainage and sewers are installed and finally 12 inches of gravel is placed on the bed followed by compaction.

The pavement is laid with concrete mixed along with the mesoporous adsorbent. The road surface is created, smoothed and trimmed off to the necessary level. After the road surface is ready to use, vehicles are allowed to travel. Due to the combustion of fuels in engine, gases are emitted. The air pollution caused is measured in this case. And the air pollution caused by the vehicles passing on normal road is also measured. Both the pollution levels are measured and compared. The effectiveness of the mesoporous adsorbent in road pavement is observed.

3.2 Summery & conclusion

From the study it can analyzed that addition of mesoporous adsorbent in the road pavement helps in capture and reduction of concentration of pollutants released from the vehicles. Effective adsorption of the vehicular emissions can be enhanced with the mesoporous adsorbent. Through the addition there is no effect on the properties such as durability, workability and strength of the pavement. The desired properties can be achieved along with the materials used in road pavement. Hence from the study addition of mesoporous adsorbent in the road pavement is suggested helpful, economical, useful, environmental and sustainable contributing to the reduction of air pollution.

So, addition of mesoporous adsorbent in road pavements results in prosperous and healthy environment with a healthy construction. All the citizens should be given awareness of the uses and promote the importance of controlling air pollution for the betterment of environmental condition

References

1. Alessandra Zanoletti, Ivano Vassura, Elisa Venturini, Matteo Monai; A New Porous Hybrid Material Derived From Silica Fume and Alginate for Sustainable Pollutants Reduction; Forontiers in Chemistry, March 2018.
2. Jong Whan Rhim; Physical and chemical properties of water resistant sodium alginate films; LWT- Food science and Technology; Volume 37, May 2004.
3. Two authors including Muhammad Arif Ab Aziz; Mesoporous adsorbent for CO2 capture application under mild condition; Journal of Environmental Chemical Engineering, March 2019.

4. Md. Shohel Reza Amin, Umma Tamima, and Luis Amador Jimenez; Understanding Air Pollution from Induced Traffic during and after the Construction of a New Highway: Case Study of Highway 25 in Montreal; October 31, 2017.
5. Ehsan Teymouri, Hojat Karami, Sayed Farhad Mousavi and Saeed Farzin; Reducing Urban Runoff Pollution Using Porous Concrete Containing Mineral Adsorbents; January 2020.
6. Taylor and Francis Ltd; Road Materials and Pavement Design.
7. Masoud Kayhanian, HuiLi, John T Harvey; Application of permeable pavements in highways for stormwater runoff management and pollution prevention; January 17, 2019.
8. Sriman Narayanan. K and Anupriya; Smog Absorbing Pavement; Volume 7, April13, 2019.
9. Scott Koch, Khaled Ksaibati and George Huntington; Performance of Recycled Asphalt Pavement in Gravel Roads; Journal of the Transportation Research Board, January 1, 2011.
10. A. Athanasopoulou and George Kollaros; Fly ash exploited in pavement layers in environmentally friendly .

Risk Assessment Prediction for Long Term Settlement using Monte Carlo Simulation for Open Dumping Area

[1]*Nur Irfah* Mohd Pauzi, [2]*Thang* Kel Win

[1] Civil Engineering Department, College of Engineering, University Tenaga Nasional Berhad (UNITEN), Putrajaya Campus, Jalan Kajang, Puchong, 43000, Kajang, Selangor,Malaysia
[2] Institute of Energy Infrastructure, College of Engineering, University Tenaga Nasional Berhad (UNITEN), Putrajaya Campus, Jalan Kajang, Puchong, 43000, Kajang, Selangor, Malaysia.

Abstract Currently, there are many abandoned open dumping areas that were left without any proper mitigation measures. Deformation properties of geo-materials are considered as the necessary information to perform sound and precise stability analysis both in overall or partial aspects. In the case landfill engineering, since Municipal Solid Waste, MSW, are the main material shaping the body of these constructions, having a proper background of the compressibility and stiffness of these materials to execute the stability analyses in design procedure is necessary. Open dumping area has a potential hazard such as the settlement of the dumping sites, the release of hazardous gas and pollution of the groundwater to the environment. The objective of this paper is to determine the risk assessment at the open dumping area after they has been closed using Monte Carlo Simulation method. The risk assessment exercise is conducted at the Kuala Lumpur dumping area. The rapid urbanisation of Kuala Lumpur coupled with increase in population lead to increase in waste generation. It leads to more dumping/landfill area in Kuala Lumpur. The first stage of this study involves the assessment of the dumping area and samples collections. It followed by measurement of settlement of dumping area using oedometer. The risk of the settlement is predicted using Monte Carlo simulation method. Monte Carlo simulation calculates the risk and the long-term settlement. The model simulation result shows that risk level of the Kuala Lumpur open dumping area ranges between Level III to Level IV i.e. between medium risk to high risk.

1 Introduction

In the current economic context, marked by globalization and by the growing interdependences between environment and development, it has become largely accepted that the environmental risks management is both an economic and a social priority. The globalization of environmental issues leads to growing concerns, especially for companies aiming at obtaining ecological performances, having the purpose to keep under control the impact on the environment of their activities, products and services (Udrea, 2011)Sustainable waste management simply means managing waste by prioritising as per the waste hierarchy .This implies waste reduction is the topmost priority if possible. The other priorities in descending order are reuse; recovery via recycling, composting, and energy; and disposal which also includes landfilling. Most of the waste produced, particularly in the UK, is generally disposed to landfills. Waste disposal to landfills, in general, is an easy and low-cost waste management option but it does raise environmental concerns. During the process of waste degradation, landfills produce waste products in three phases (Fig. 1). These are solid (i.e., degraded waste); liquid (i.e., leachate, which is water polluted with wastes); and gas (usually referred to as landfill gas).

Further, landfills and their aforesaid waste products may pollute the three principal environmental media –the atmosphere, the lithosphere and the hydrosphere (Fig. 2). Such pollution will be transmitted through these media and will have an impact, either directly or indirectly, upon human, the natural environment (including aquatic and terrestrial flora and fauna) and the built environment. This highly necessitates that hazards and risks of landfills be assessed and managed to guard the environment and its species from landfill hazards.

2 Sustainable Development Goals

Malaysia together with other 192 world leaders adopted the 2030 Agenda for Sustainable Development (2030 Agenda) at the United Nations General Assembly in New York on 25 September 2015. This is a global commitment towards a more sustainable, resilient and inclusive development, with 17 Sustainable Development Goals (SDGs) and 169 targets. The SDGs is a new, universal set of goals, targets, and indicators that UN member state will be expected to use to frame their agenda and political policies over the next 15 years (2016 – 2030).

© The Authors, published by EDP Sciences. This is an open access article distributed under the terms of the Creative Commons Attribution License 4.0 (https://creativecommons.org/licenses/by/4.0/).

With its 17 Goals, 169 Targets and more than 200 Indicators that cover 5 dimensions namely People, Planet, Prosperity, Peace and Partnership, the SDGs will stimulate action over the next fifteen years in areas of critical importance for humanity and the planet.

In the process to embrace and implement the 17 SDGs in a systematic and measurable manner, Malaysia has put in place an enabling environment through the following initiatives:

- established a multi-stakeholder, participatory governance structure helmed by the National SDG Council chaired by the Prime Minister;
- held several national SDG symposiums and focus group sessions to promote awareness and participation of stakeholders;
- conducted studies on data readiness and gap analysis;
- undertaken a mapping exercise involving nongovernment and civil society organisations and the private sector to align SDGs with 11MP initiatives;
- formulated a National SDG Roadmap to guide implementation of the 2030 Agenda and the SDGs; and
- implementing SDG initiative under the framework of the 11MP. Moving forward, Malaysia will implement the 2030 Agenda under the framework of 11MP, whereby the means of implementation include:
- localising SDGs at sub-national levels by replicating the national multi-stakeholder governance structure at state levels.
- mobilising resources and funding through partnerships with stakeholders including crowd sourcing, social entrepreneurship and Corporate Social Responsibility programmes besides public sector funding under 11MP; and
- strengthening data readiness and filling data gaps to develop a comprehensive dataset for SDG implementation.

3 Monte Carlo Simulation

Monte Carlo simulation is a statistical technique by which a quantity is calculated repeatedly, using randomly selected "what-if" scenarios for each calculation. Though the simulation process is internally complex, commercial computer software performs the calculations as a single operation, presenting results in simple graphs and tables. These results approximate the full range of possible outcomes, and the likelihood of each. When Monte Carlo simulation is applied to risk assessment, risk appears as a frequency distribution graph similar to the familiar bell-shaped curve, which non-statisticians can understand intuitively. The technique is used by professionals in such widely disparate fields as finance, project management, energy, manufacturing, engineering, research and development, insurance, oil & gas, transportation, and the environment Monte Carlo simulation also has important limitations, which have restrained EPA from accepting it as a preferred risk assessment tool:

1. Available software cannot distinguish between variability and uncertainty. Some factors, such as body weight and tap water ingestion, show well-described differences among individuals. These differences are called "variability". Other factors, such as frequency and duration of trespassing, are simply unknown. This lack of knowledge is called "uncertainty". Current Monte Carlo software treats uncertainty as if it were variability, which may produce misleading results.

2. Ignoring correlations among exposure variables can bias Monte Carlo calculations. However, information on possible correlations is seldom available.

3. Exposure factors developed from short-term studies with large populations may not accurately represent long-term conditions in small populations.

4. The tails of Monte Carlo risk distributions, which are of greatest regulatory interest, are very sensitive to the shape of the input distributions.

4 Summary of Literature Review

Table 1. Summary of Literature Review

Item	Title	Author	Findings
1	Risk Assessment Predictions of Open Dumping Area after Closure using Monte Carlo Simulation	Nur Irfah Mohd Pauzi, Mohd Shahril Mat Radhi, Husaini Omar	-In conclusion, the laboratory testing of the samples show the settlement for 3 kg, 5kg, 7 kg and 9kg loadings are 2.15 mm, 2.35 mm, 2.55mm and 2.50 mm respectively. It shows that higher applied load will produce higher displacement regardless of the waste types found in the soil sample. -The proposed remedial work is using bore pile or micropile. -The output from this risk assessment analysis implied that Monte Carlo Simulation model can be implemented in Malaysia especially in Kuala Lumpur dumping areas.
2	Monte Carlo Simulation in Environmental Risk Assessment-- Science, Policy and Legal Issues	Susan R. Poulter	- Monte Carlo simulation and other methods of probabilistic risk assessment are already bringing more information and more sophisticated uncertainty analysis into risk assessment. - If Monte Carlo and other probabilistic methods are to fulfil their promise in environmental risk assessment, the discussion of those issues and the development of appropriate guidance and legal rules must grow apace.
3	Use of Monte Carlo Methods in Environmental Risk Assessment at the INEL: Applications and Issues	Gerald Harris, Robin Van Horn	- The slope factor was only the third most important input parameter; it generally would be expected to be of much greater importance. They derived these distributions by adjusting the EPA recommended value for a variety of factors which are used to derive the slope factor. - The averaging time distribution, which was based on age at death, did not have a significant impact on risk. This distribution was slightly conservative; the median averaging time was about 73 years, while the true median is closer to 78 years - 5. The analyst would indicate the best estimate is 99% of the 54 population would be exposed to risks of less than 1 x 10*6 . The analyst would explain that the relatively high point estimates (1 x 10"5) was mostly due to the use of the maximum benzene concentration in the soil. This occurred because of the great amount of variation in the observed benzene concentrations.
4	Monte Carlo method application for environmental risks impact assessment in investment projects	Marieta Olaru, Mircea Şandru, Ionela Carmen Pirnea	- The use of the Monte Carlo simulation method to assess the risk impact in general and for the environmental risks in particular, taking as an example a Romanian energy investment project allowed to establish the uncertainty level in

			attaining the planned key-performance indicators for the respective project - The result was an increase of the chances to meet such indicators with optimal effort, adjusted on the basis of the multi-iterative analysis carried out using the Monte Carlo method. -The use of this method allowed establishing the risks and the impact of such risks, especially with regard to: the environmental impact of the processes of the project, the impact of various stress factors on the technical economic processes of the project, especially the processes that ensure the health and occupational safety of the project teams.
5	Monte Carlo method application for environmental risks impact assessment in investment projects	Marieta Olaru, Mircea Şandru, Ionela Carmen Pirnea	- The use of the Monte Carlo simulation method to assess the risk impact in general and for the environmental risks in particular, taking as an example a Romanian energy investment project allowed to establish the uncertainty level in attaining the planned key-performance indicators for the respective project. - . The result was an increase of the chances to meet such indicators with optimal effort, adjusted on the basis of the multi-iterative analysis carried out using the Monte Carlo method. -The use of this method allowed establishing the risks and the impact of such risks, especially with regard to: the environmental impact of the processes of the project, the impact of various stress factors on the technical economic processes of the project, especially the processes that ensure the health and occupational safety of the project teams.
6	The use of the Monte Carlo method for predicting environmental risk in construction zones	E Smirnova	- Thus, the concept of ERF means the relationship between territories that include a certain number of natural components and characterize them as "risk situations". - When choosing a model of environmental risk in order to predict the level of interaction of the construction object with the environment, it is necessary to take into account the features of natural landscapes. -The adjusted matrix of the distribution of construction zones and ecosystem vulnerability risk based on the study of the parameters of anthropogenic impact on the components of the environment allowed us to correlate the index of regions (regions) and the index of ecological groups for natural landscapes.
7	Risk Assessment Models to Improve Environmental Safety	Arkadiy Larionov, Ekaterina Nezhnikova, Elena Smirnova	- The article analysed the applicability of the Monte Carlo method to the optimization of environmental safety in the field of construction,

	in the Field of the Economy and Organization of Construction: A Case Study of Russia		housing, and communal services. Research of this kind is one of the first (judging by the review of the available literature). -The objective of the study was to develop a method for the construction and utility sector and create a methodology that makes it possible to predict, depending on the impact of anthropogenic and natural factors, how a building or structure will be built and operated at all stages of its life cycle

5 Results and Discussions

5.1 Settlement of Waste Soil

The oedometer test shows that the settlement of the waste soil are in the range of 2.15 mm to 2.55 mm. For 9kg loading, the waste soil displaced at 2.5 mm, similarly at 7 kg loading, the waste soil displaced at 2.55 mm and at 5 kg loading, the waste soil displaced at 2.35 mm and finally for 3 kg loading, the soil displaced at 2.15 mm. The displacement become constant once it reaches 100 percent degree of consolidation. Fig 1 shows the displacement versus the square root of time for waste soil. The settlements are further calculated using the other model such as Soil Mechanic Based model, Bjarngard and Edgers model, power creep function hyperbolic function and rheological model. The calculations are based on the settlement of open dumping area.

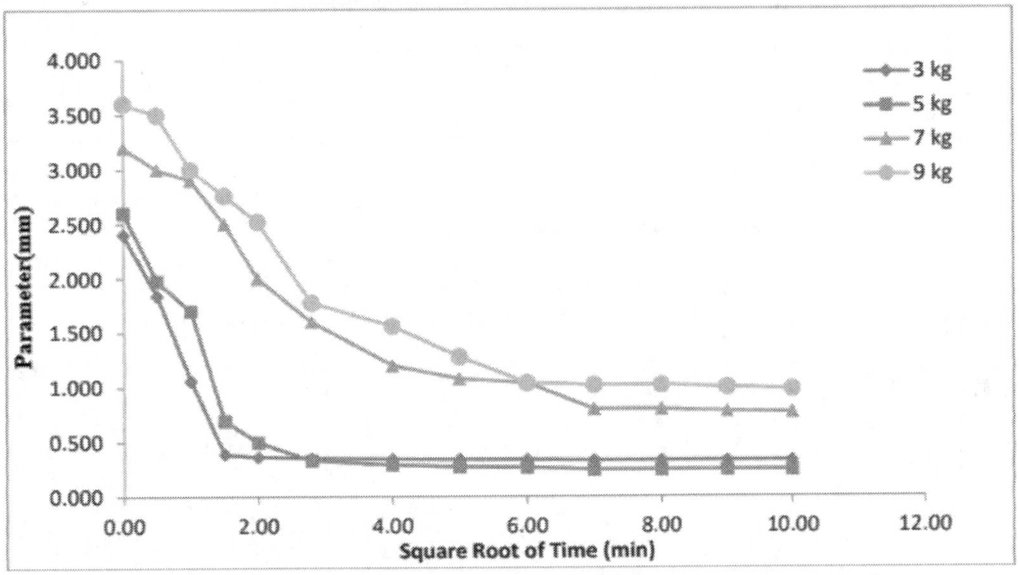

Fig.1. Displacement versus square root of time for waste soil at Kuala Lumpur open dumping area

The highest settlement found at the open dumping area is 11.6 meter and the lowest is 0.0151 meter as shown in Table 1 based on the Monte Carlo predicted settlement. Case (1) shows the result based on calculation of the settlement with the duration of settlement for primary settlement which is 25 days. On the other hand, Case (2) calculated the settlement with the duration of settlement for secondary settlement that occur at 25 years. The analysis of settlement is then evaluated based on its risk level. The depth of the settlement will determine the risk level.

Table 2. Settlement Calculations using Monte Carlo simulation prediction and Excel Spreadsheet

Settlement Models	Monte Carlo Predicted Settlement (m)	Excel Spreadsheet calculation of Settlement (m)
SMBM	Case (1) : 0.0907 Case (2): 11.662	Case (1): 1.139 Case (2): 24.234
BEM	Case (1): 7.115 Case (2): 3.547	Case (1): 6.131 Case (2): 4.663
PCF	Case (1): 0.0151 Case (2): 0.1392	Case (1) : 0.042 Case (2): 0.13
HF	Case (1): 0.0151 Case (2): 0.1986	Case (1): 0.029 Case (2): 0.1987
RM	Case (1): 1.525 Case (2): 2.253	Case (1): 0.426 Case (2): 0.882

Note: SMBM – Soil Mechanics Based Model, BEM – Bjarngard and Edgers Model, PCF – Power Creep Function, HF – Hyperbolic Function, RM – Rheological Model

5.2 Settlement Evaluation Statement and Risk Level

Settlement evaluation statement is a statement that used to evaluate the potential to settle of the dumping area. The statement are either YES or NO. If the dumping area is potential to settle, then potential to settle is YES. If the dumping area has no indication to settle then the potential to settle is NO. This settlement evaluation is done after the analysis by Monte Carlo Simulation and statistical analysis. Table 2 tabulates the settlement evaluation statement and the level of risk for dumping area.

Table 3 shows the risk assessment result for Kuala Lumpur dumping area. The analysis is based on the five settlement models. Result shows that Case 2 have a higher risk as compared to Case 1. The potential for the Kuala Lumpur dumping area to settle are based on its settlement depth. From the summary of the data in Table 3, it shows that the dumping area has the risk Level IV: High Risk. As for Case 1, the risk is level I: No Risk and it is not potential to settle. Thus, no remedial action is required. The remedial work should be made after 25 years.

Table 4 shows the suggestion work for settlement of waste soil at open dumping area. The level IV with the total settlement of more than 1 meter would suggest the settlement of the waste soil may get deeper and possibility of reaching more than 5 meters. The usage of bored pile or micro pile to the foundation level below the debris dumping strata or the used of prefabricated vertical drain (PVD) are suggested as remedial action for the dumping area.

Table 3. Settlement Evaluation Statement and Level of Risk

Level	Description	Settlement Value	Potential to Settle Statement
I	No Risk (NR)	Less than 0.05 m	NO or YES
II	Low Risk (LR)	0.05 m to 0.10 m	YES
III	Medium Risk (MR)	0.10 m to 1.0 m	YES
IV	High Risk (HR)	More than 1.0 m	YES

Table 4. Risk Assessment at Kuala Lumpur Dumping Area

Settlement Models	Monte Carlo Predicted Settlement (m)	Risk Assessment Level	Potential to Settle/ Settlement to Occur
SMBM	Case (1) : 0.0907 Case (2): 11.662	Case (1) Level I Case (2): Level IV	No Yes
BEM	Case (1): 7.115 Case (2): 3.547	Case (1): Level IV Case (2): Level IV	Yes Yes
PCF	Case (1): 0.0151 Case (2): 0.1392	Case (1): Level I Case(2): Level III	No Yes
HF	Case (1): 0.0151 Case (2): 0.1986	Case (1) : Level I Case (2) : Level III	No Yes
RM	Case (1): 1.525 Case (2): 2.253	Case (1): Level IV Case (2) : Level IV	Yes Yes

Table 5. Suggestion Work for settlement of Waste Soil at Open Dumping Area

Level	Settlement (m)	Suggestion Work
I	Less than 0.05	• None
II	0.05 to 0.10	• Check drainage, drain out the water ponding on the waste soil surface
III	0.10 to 1.0	• Remove the unsuitable or problematic soil, fill the area with good soil • Drain out the water ponding on the waste soil surface
IV	More than 1.0	• The settlement of the waste soil may get thicker which could be more than 5 meters, used bored pile or micropile drilled to the foundation level below the debris dumping strata OR used prefabricated vertical drain (PVD)

5.3 Confined Compression Tests

Confined compression tests were performed in an oedometer with nominal dimensions of 497 mm height and 548 mm diameter. Fig 2 presents the results of 4 typical compression curves. Three fresh waste and one 4 years old samples were used. It can be noted that not only the primary compression index but also the swelling index seem to be decreased with the waste age. The compression indexes in the case of the fresh waste samples are similar; however, the rebound indexes are significantly different.

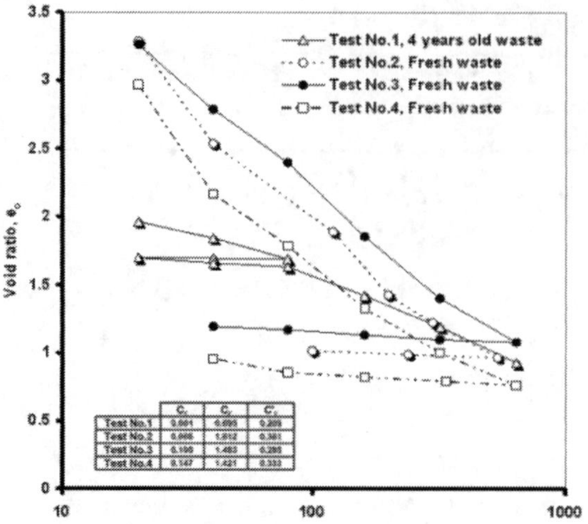

Fig 2. Confined compression curves achieved from different MSW samples.

Fig 3 compares the obtained results to the primary compression index with others published in the technical literature.

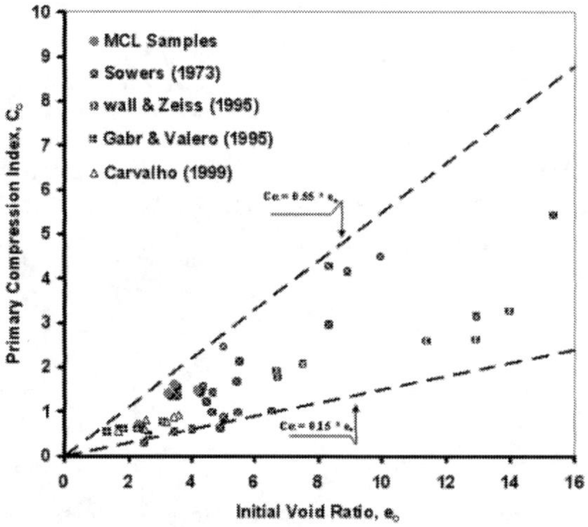

Fig 3. Measured primary compression index comparing reported values in literature.

As can be observed, the results obtained using the MCL waste samples fit in the range suggested by Sowers (1973) and are also compatible with the results presented by Gabr and Valero (1995). According to Landva et al. (1984), the value of the normalized primary compression index, C'c, varies from 0.2 to 0.5 which agrees with the results achieved in this study. Estimation of constrained modulus of MSW is another way to estimate the waste fill primary settlement. Waste placement can be a one-dimensional compression problem (e.g., waste is placed over a large area in relation to the thickness of the deposit). An increment of vertical effective stress, v, produces an increase in vertical strain, εv. A constrained modulus D, can be defined as follow:

$$D = \frac{\Delta \sigma'_v}{\Delta \varepsilon_v} \tag{1}$$

Using this parameter and based elastic theory, elasticity modulus of MSW also might be calculated as follow:

$$E = \frac{D(1+v)(1-2v)}{(1-v)} \tag{2}$$

Where, v is Poisson's ratio. The MSW constrained modulus appears to be linearly dependent of the applied vertical stress and its values varied from 100 to 5500 kPa, depending on the value of vertical stress.

6 Conclusions

In conclusion, the laboratory testing of the samples show the settlement for 3 kg, 5kg, 7 kg and 9kg loadings are 2.15 mm, 2.35 mm, 2.55mm and 2.50 mm respectively. It shows that higher applied load will produce higher displacement regardless of the waste types found in the soil sample. Further analysis using settlement model and Monte Carlo simulation shows that the settlement is at 11.6 meters based on soil mechanic model. The risk evaluation based on the settlement show that the risk level for this area is at Level IV i.e. high risk. The proposed remedial work is using bore pile or micro pile. The output from this risk assessment analysis implied that Monte Carlo Simulation model can be implemented for open dumping area. Moreover, the results showed that normalized long-term compression index could be 10 times higher than short-term compression indexes since the effect of biodegradation is more pronounced in long terms. This can be proved from the results of manual monitoring at the field with the calculated settlement model.

Landfills continue to be one of the main methods of waste disposal despite their relatively high potential to pollute the environment. Therefore, risk assessment is required as a tool to identify and define landfill hazards for the environment. The risk assessment is a most important factor of an effective risk control, as the degree of success of the latter is based on the former. This research allows the authors to recognise the necessity, identify knowledge gaps and establish bases for developing a more holistic framework of an algorithmic and quantitative methodology of risk analysis for landfill leachate in an integrated manner. With the success of this finding, Malaysia will a step closer to the aims of sustainable development goals which is to achieve the sustainable cities and communities. This action will indirectly increase the reputation of our country and prove that Malaysia had the ability to reuse all the waste materials in the construction field. Thus, this can be said as kill two birds with one stone because in the same time Malaysia increased its technology in construction while reducing the waste material which is a difficulty faced by every country.

7 Acknowledgment

The author would like to acknowledge the University Tenaga Nasional for supporting the research under the TNB Seed Fund BOLD grant J510050002/2022028 and also to the Civil Engineering Department, UNITEN for providing the facilities for library resources.

8 References

1. Abah S.O. and Ohimain E.I., World Applied Sciences Journal 8(4), 436-442 (2010)
2. Kurian J., Esakku S., Nagendran R. and Visvanathan C., A decision making tool for dumpsite rehabilitation in developing countries, Proceeding Sardina 2005, Tenth International Waste Management and Landfill Symposium, Caglian, Italy, (2005)
3. Nur Irfah M.P., Electronic Journal Geotechnical Engineering 17, (2012)
4. Sivakumar Babu G.L., Lakshmikanthan P. and Santhosh, Assessment of Landfill Sustainability, In: Sivakumar Babu G., Saride S., Basha B. (eds) Sustainability Issues in Civil Engineering. Springer Transactions in Civil and Environmental Engineering. Springer, Singapore, (2017)

5. Sirsat M.S., Cernadas E., Fernandez- Delgado and R.Khan, 2017 Classification of Agricultural Soil Parameters in India, Comput. Electron. Aric., Vol 135, pp. 269 – 279.
6. Lakshmikanthan P., Santhosh L.G. and Sivakumar Babu G.L., Evaluation of Bioreactor Landfill as Sustainable Land Disposal Method, In: Sivakumar Babu G., Saride S., Basha B. (eds) Sustainability Issues in Civil Engineering. Springer Transactions in Civil and Environmental Engineering. Springer, Singapore, (2017)
7. Yakubova G., Kavetskiy A., Prior S.A. and Torbert H.A., 2016 Applying Monte Carlo Simulation to optimize an in elastic neutron scattering system for soil carbon analysis, Appl. Radiat. Isot., vol. 128, pp. 237-248.
8. Sealy K.S. and Strobl E., 2017 A hurricane loss risk assessment of coastal properties in the Carribean: Evidence from the Bahamas, Ocean Coastal Management, vol. 149, pp. 42-51.
9. Abah S.O. and Ohimain E.I., World Applied Sciences Journal 8(4), 436-442 (2010)
10. Kurian J., Esakku S., Nagendran R. and Visvanathan C., A decision making tool for dumpsite rehabilitation in developing countries, Proceeding Sardina 2005, Tenth International Waste Management and Landfill Symposium, Caglian, Italy, (2005)
11. Nur Irfah M.P., Electronic Journal Geotechnical Engineering 17, (2012)
12. Sivakumar Babu G.L., Lakshmikanthan P. and Santhosh, Assessment of Landfill Sustainability, In: Sivakumar Babu G., Saride S., Basha B. (eds) Sustainability Issues in Civil Engineering. Springer Transactions in Civil and Environmental Engineering. Springer, Singapore, (2017)
13. Lakshmikanthan P., Santhosh L.G. and Sivakumar Babu G.L., Evaluation of Bioreactor Landfill as Sustainable Land Disposal Method, In: Sivakumar Babu G., Saride S., Basha B. (eds) Sustainability Issues in Civil Engineering. Springer Transactions in Civil and Environmental Engineering. Springer, Singapore, (2017)
14. EPA, 2000. Risk Characterization Handbook, (U.S. Environmental Protection Agency, Office of Science Administration, Washington, DC).
15. EPA, 1991. Standard Default Exposure Factors, Risk Assessment Guidance for Superfund, Volume I: Human Health Evaluation Manual Supplemental Guidance, (U.S. Environmental Protection Agency Office of Solid Waste and Emergency Response, Toxics Integration Branch, Washington, DC, OSWER Directive 9285:6-03).
16. Smith, R.L. In press. Use of Monte Carlo simulation for human exposure at a Superfund site. Submitted to Risk Analysis, May 1993.
17. Springett, D. (2003). Business conceptions of sustainable development: A perception from critical theory. Business Strategy and the Environment, 12, 2, 71–86, in Laine, M. (2005). Meanings of the term 'sustainable development' in Finnish corporate disclosures. Accounting Forum, 29, 395-413.

Assessment of air quality and damage caused after an anthropogenic disaster in the city of Beirut.

Jayalakshmi. S[1] *Jayanthi.* B[2] , Ezhilarasi R[3]
1 Professor, Department of Civil Engineering, College of Engineering Guindy, Chennai, India – 600025; Email: j_lak2001@yahoo.co.in
2 Ph.D. Research Scholar, Department of Architecture and Planning, College of Engineering Guindy, Chennai, India – 600025; Email: jaya.march.021@gmail.com
3 Pg student, Department of Civil Engineering, College of Engineering Guindy, Chennai, India – 600025

Abstract. Air pollution is an important cause of concern and has a significant impact on health as well as the environment. Air pollutants are released into the atmosphere by various sources, and the concentration in cities is mainly due to anthropogenic sources like emissions from industries, factories, vehicles, fossil fuel combustion, waste production, etc. But certain high episodes of air pollution can be witnessed in cities which can cause serious threats to human beings and the environment. Remote sensing is one of the sustainable measures, which allows the integration and presentation of useful information at various temporal and spatial scales. This tool can be helpful in understanding and managing natural disasters and man-made risks contributing to air pollution. Analyzing the cause of such high pollution episodes can help to take necessary steps to reduce the emissions, and damage or to take precautions to avoid man-made disasters. One such incident of an explosion of ammonium nitrate in a port had a long-lasting impact on the air quality in Beirut where the average concentration of air pollution already exceeds WHO threshold limits. This paper explores the concentration of aerosols and NO_2, before and after the incident of an explosion that took place in Beirut. Assessment of the damage near the port caused by the explosion in Beirut was done using a change detection technique for which an algorithm was developed. Due to the non-availability of field data for the study area of Beirut, the algorithm was developed and validated for Chennai city and extended to Beirut. The aerosol and NO_2 concentration data were extracted from sentinel-5p/TROPOMOI and validated with ground data through regression analysis. The results obtained from this study can be used as an early warning system and decision support system for mitigating loss due to the explosion.

Keywords: air pollution, aerosol, NO_2, satellite data, damage assessment

1. Introduction.

Air pollution is a major environmental risk factor for poor health and causes about 2 million premature deaths worldwide each year [1]. Natural and anthropogenic activities are major sources of air pollution. There are 6 major criteria pollutants out of which the exposure to particulate matter (PM2.5) and nitrogen oxides (NOx) shows severe damage to health as well as often exceeding the World Health Organization (WHO) guidelines threshold limit [2,17]. Regular exposure to NO_2 and aerosol has been linked to an increased risk of respiratory problems, cardiovascular disease along with lung cancer. It can also cause a myriad of short-term health issues such as respiratory infections, irritation of the eyes, nose, and headaches. In addition to health concerns, suspended particulates also reduce visibility

© The Authors, published by EDP Sciences. This is an open access article distributed under the terms of the Creative Commons Attribution License 4.0 (https://creativecommons.org/licenses/by/4.0/).

and contain harmful compounds which can cause environmental damage via dry or wet deposition.

There are different sources that add pollutants into the atmosphere reducing the air quality. Apart from these at times disasters or incidents that happen can leave a remarkable impact on air pollution [2]. The impact of fire smoke in the boreal forest showed an impact on the air quality in the US [3]. One such incident of explosion took place in Beirut, increasing the concentration of NO_2 and aerosol to an alarming level.

On August 4, 2020, nearly 3,000 tons of the highly combustible chemical compound (ammonium nitrate) exploded in Beirut's port [4], shooting a plume of orange smoke into the sky, followed quickly by a pressure wave that ripped through nearby buildings and blew out windows across the metropolis. In ordinary circumstances, ammonium nitrate is a fertilizer, but when placed under the stress of heat and pressure, it acts like a bomb. When ammonium nitrate explodes it releases toxic gases like nitrogen oxide (NO_x) and ammonia gas and also emits particulate matter (PM). These gases, both harmful to the human respiratory system and the environment, in turn, break down and react with other chemicals. Nitrogen oxides, for example, when combined with other pollutants and sunlight form "bad ozone," a secondary pollutant, ozone at ground level. Regular monitoring and reducing the amounts of air pollution is therefore critical in this region.

Satellite remote sensing has a wide range of applications and one such application is retrieving data on air quality [5]. These techniques can be used to study spatial distribution, temporal variations, fine and coarse mode fraction, vertical distribution, etc. to derive sources, transportation, and interaction between pollutants and clouds [6,7,8]. The most significant factor is its ability to fill in gaps where ever the ground data information on air quality is the unavailability of sufficient monitoring stations. There are several air pollution retrieving satellites like MODIS, OMI, MISR, TOMS, AVHRR, MERIS, AATSR, etc. and algorithms are available. The improved resolution of TROPOMOI is expected to reduce the effect of spatial averaging compared to OMI and other coarse-resolution satellites [9]. This study aims to analyze the intensity of aerosol and NO_2 concentration from sentinel-5p/TROPOMOI and validate it. Also, to understand the impact of a disaster on the urban structure and human health.

2. Materials and Methods

The study explores the concentration levels of aerosol and NO_2 before and after the explosion in Beirut. Since the ground-based data was unavailable for Beirut, the satellite data and ground-based data from the Central Pollution Control Board (CPCB) for the city of Chennai were used for validation. Firstly, the ground-based data for PM 2.5, PM 10, and NO_2 were collected for the dates 12, 13, and 14th Jan 2021. The data was sourced from CPCB for the Chennai region. Then the satellite images of the Chennai region coinciding with the same dates were extracted from the Sentinel-5 Precursor mission instrument which collects data useful for assessing air quality. It uses the TROPOMI instrument which is a multispectral sensor that records the reflectance of wavelengths important for measuring atmospheric concentrations of pollutants. S5P is a near-polar sun-synchronous orbit satellite flying at an altitude of 817 km, with a swath width of approx. 2600 km, an along-track resolution of 7km, and daily global coverage [10]. The day-to-day variability of the data obtained from satellite and in situ measurements was found to be similar [9]. Here the data obtained are in the form of vertical column densities which are measured as mol/m^3. Using a look-up table of altitude-dependent air-mass factors (AMFs) the vertical column densities are converted before which the tropospheric and stratospheric slant column densities are separated from the total slant column using a data assimilation system based on the TM5-MP chemical transport model [9]. NO_2 measurements are available from the CPCB measurement database as hourly averaged concentrations in units of μgm^{-3} for Chennai city. Regression analysis was used to evaluate the extent to which the satellite data can resolve ground-based estimates [11]. It involves analyzing the goodness of fit of the regression. One measure of goodness of fit is the coefficient of determination (R^2). Both satellite data and metrological data were combined to derive the correlation of NO_2, and AOD [12]. The integration of surface, satellite, and meteorological products and these products are added to form multiple linear regression (MVM) equations to estimate mass concentration [6].

Secondly, the concentration maps of Aerosol and NO_2 before and after the explosion of Beirut were prepared and analyzed. QGIS software was used for the visualization of maps. Thirdly damage assessment was done to identify the intensity of damage through a change detection technique. The satellite images obtained were processed and analyzed using the SNAP toolbox. To improve the accuracy of the results, both sentinel-1 (Radar data) and sentinel-2 (optical) data were used for pre-processing involving two major steps - Applying orbital file details and back geocoding. Then Terrain correction was carried out to correct the geometric distortions that lead to geolocation errors. It is done to move the image pixels into the proper spatial relationship with each other. The process of combining multiple images into a single image was done by layer stacking for which the images were re-sampled to have the same spatial resolution. Finally, the Change detection technique is used to estimate the damages caused by the explosion for which the algorithm was developed. The damages caused by the explosion are estimated by the change detection technique. The framework of the methodology followed in the study is given below in Fig. 1

3. Study Area and Data Source

3.1 Study Area

Beirut is the capital and largest city of Lebanon. The latitude and longitude of Beirut, Lebanon 33° 53' 19.0680" N and 35° 29' 43.7280" E. Beirut is one of the most religiously, culturally, and economically diverse cities in the Middle East. Beirut is the third-largest city in the Levant region and the fifteenth-largest in the Arab world. Beirut covers an area of 19.8 km^2 and its metropolitan area is 67 km^2. Beirut is comprised 60% of the urban population of Lebanon. Beirut's Population is almost five times the population of the second largest city in Lebanon, Tripoli. Now, it is estimated that the population has reached 2 million. Beirut has a warm climate with wet winters. The average temperature estimated in Beirut was 20.5 °C. The annual rainfall of Beirut is 845 mm per year. The precipitation difference between the driest and the wettest month is 196 mm. The driest and warmest months in Beirut in July and August respectively. The closest ocean to Beirut is the Mediterranean Sea. Fig 2 shows the location of Beirut city.

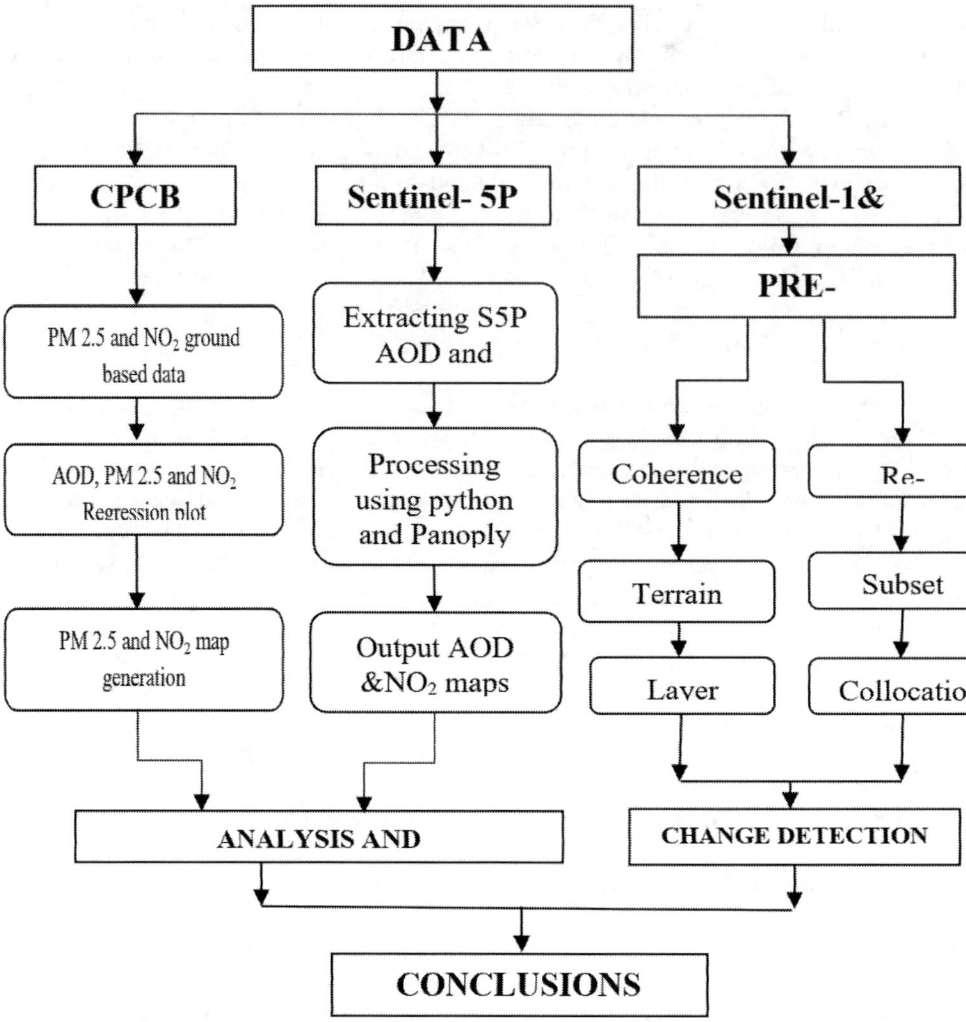

Fig 1: Methodology

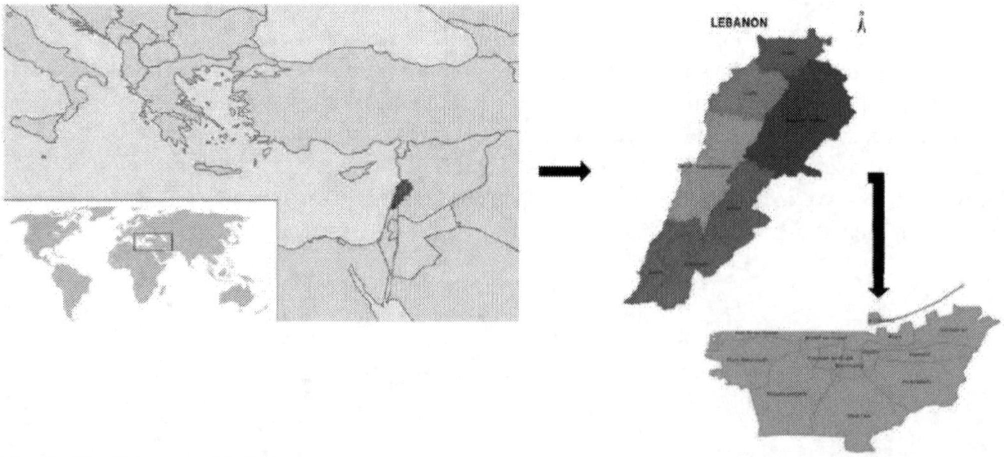

Fig 2: Study area: Beirut

3.2 Data Source

The Aerosol optical depth (AOD) data from Sentinel- 5p/TROPOMOI and NO_2 satellite data at two different wavelengths (388- 354 nm, 388-340 nm) was studied on the 13th and 14th Jan 2021 for the Chennai region. The data was processed and maps were generated in panoply software giving a quantitative estimate of the concentration of pollutants present in a vertical atmospheric column from the ground to the top of the Atmosphere (TOA).

The values extracted are given as aerosol index (AI) and the graph was plotted against AI and scan line. The NO_2 concentration from the sentinel 5P was extracted through Python script for the dates 13th to 14th Jan 2021 for the Chennai region. The values extracted are given as mol/m³ and the base map is generated using the matplotlib tool kit in Python. The quality of satellite observations was taken care of by "quality assurance value" (qa_Value) which provided an easy filter to remove less accurate observations. This study uses the Sentinel 1A, 1B and Sentinel 2A, 2B images for damage assessment. The images were downloaded from the ESA Copernicus SCI hub and necessary corrections are applied.

4. Selection of Input Parameters

4.1 Validation of data

The ground-based concentration of PM 2.5 and NO_2 was acquired from CPCB, Chennai for the dates 12th to 14th Jan. The concentration was found to be higher on 13th January 2020 as compared to the other two days. To check the accuracy of the result, a regression graph was plotted with the atmosphere (VCD) NO_2 data on the Y axis and ground level NO_2 on the X axis. Similarly, another regression graph was plotted with Aerosol Index (AI) on the Y axis and ground level PM 2.5 on the X axis. The results indicate the satellite-derived data can be a good substitute for the concentration levels acquired from the ground. The correlation graph for both pollutants has been shown in Fig 5 and Fig 6.

4.2 Analysis and Interpretation of Aerosol and NO2 in Beirut

The AOD and NO_2 concentration data from sentinel 5P at two different wavelengths (388- 354 nm, 388-340 nm) was extracted for the dates 1st to 3rd August 2020 before the explosion. The explosion occurred on the $4^{th}$of August 2020 evening at 6.00 pm. The AOD and NO_2 concentration map generated data from the 5th to 7^{h} of August 2020. About a month after another explosion and fire broke out. Similarly, the concentration maps over the period of other two consecutive months of both pollutants from 9th to 12th September 2020 and 4th to 11th November 2020 were prepared for analysis. The maps showing the major difference in concentration are shown in Fig 7 and Fig 8.

4.3 Damage Assessment in Beirut

For damage assessment, two sets of images before and after the explosion from sentinel 1 and sentinel 2 satellite was used for which a series of processing steps were applied with a snap toolbox. It estimates a constant range offset for each burst using a small block of data in the center of the burst and then it estimates a constant azimuth offset. Finally, the average of all the bursts is estimated to get the final constant range and azimuth offset. Co-registration of the product is created by keeping the same graph for which the coherence was estimated. Coherence is the process of estimating how well "correlated" the pixels are between the master and slave images used. Using the Coherence operator, the correlation between pixels is estimated. After estimating coherence in both images, it is stacked in order to be allowed to apply the Change Detection. The final step is to detect the changes between the two coherence images, and the result shows the damage caused due to the explosion. The products are re-projected so that we can insert them in QGIS for visualization and further results. Both the products are overlaid to identify what are all the buildings that were damaged due to the explosion in the Beirut port which is shown in Fig 9.

5. Results and Discussions

Before the assessment of the damage caused by the explosion, the satellite data was validated with ground-based data for the city of Chennai. The AOD concentration maps generated for the dates 13th and 14th Jan 2021 show the variation in values from -6.4 to 3.4, where the concentration was observed high on 13th Jan 2021. Also, the NO_2 concentration map generated for the dates 13th, and 14th Jan 2021 showed that the concentration level was high on 13th Jan 2021. This could be mainly due to the Bogie festival being celebrated where old things are burnt as a part of the ritual [13]. The concentration of NO_2 was higher as compared to Aerosol due to the festival ritual.

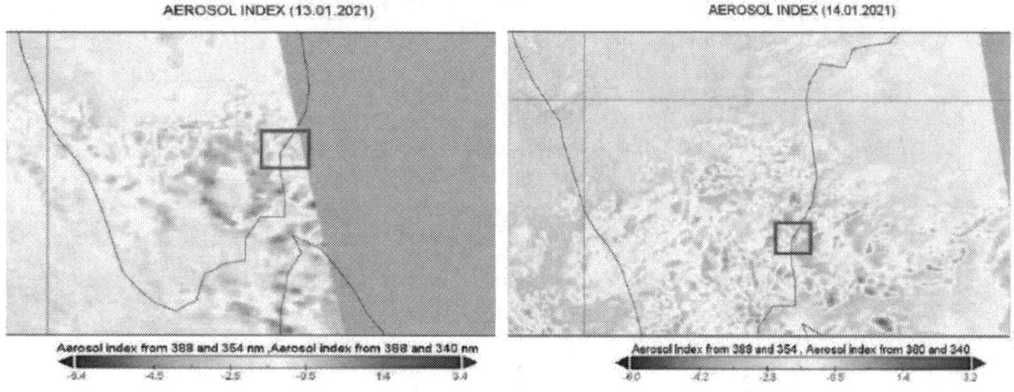

Fig 3 : Aerosol index map for Chennai region on 13th and 14th Jan 2021

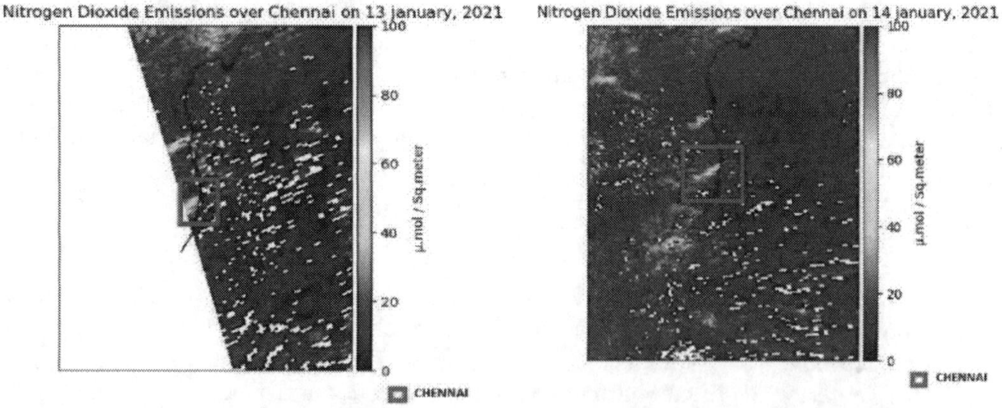

Fig 4 : NO_2 concentration map for Chennai region on 13th and 14th Jan 2021

To confirm the satellite data can be used as a good substitute for the ground-based data regression graph was plotted for both pollutants for the same dates of ground-based data. The graph for the NO_2 showed linear regression with a co-efficient of 0.806 and the AOD linear regression with a coefficient of 0.774. It indicated an excellent correlation between satellite and ground-based data [14,15].

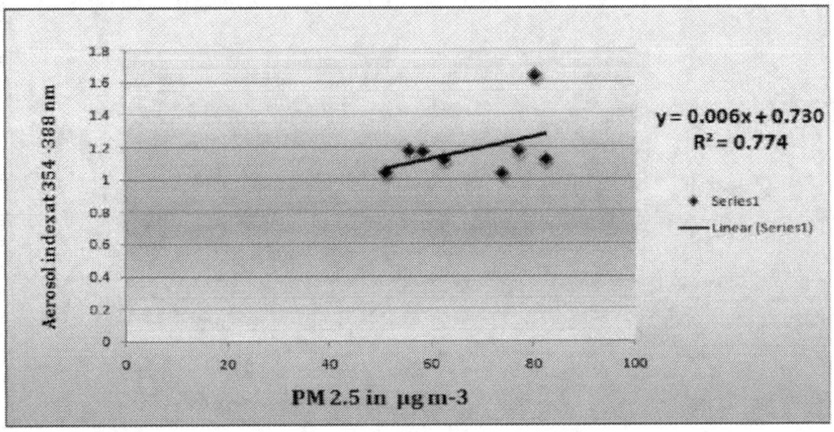

Fig 5 : Aerosol index at 354-388 nm Vs PM 2.5 in mg/m3

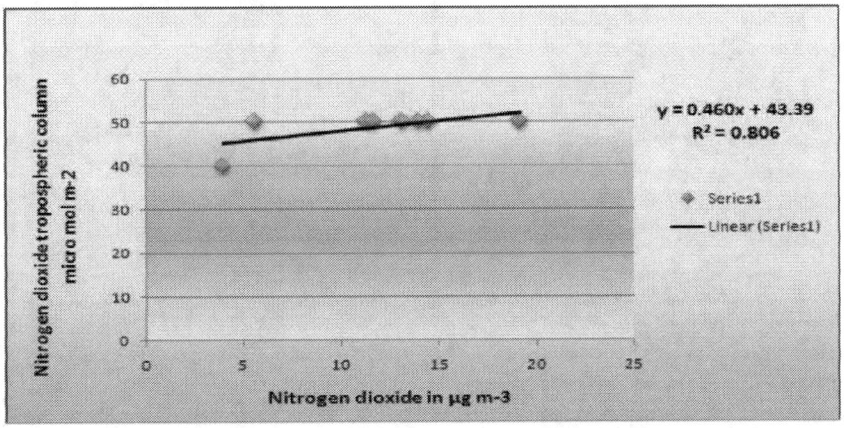

Fig 6 : Tropospheric NO2 column Vs NO_2 in mg/m3

Two pollution episodes, one before the explosion (1st -3rd August 2020) and the other one after the explosion (4th August- 7th August 2020) were considered. The AOD and NO_2 concentration maps that were generated clearly indicated the drastic increase in pollutant levels after the explosion. The aerosol concentration observed to be is less and the maximum aerosol index value is 2.3 before the explosion. The concentration map generated for the consequent months of September and November 2020 showed the impact of air quality in Beirut. The explosion had a significant reduction in air quality in Beirut between 6th August to 11th September 2020. The next consecutive month showed a gradual reduction in the levels indicating a period of at least 3 months for the pollutants to disperse.

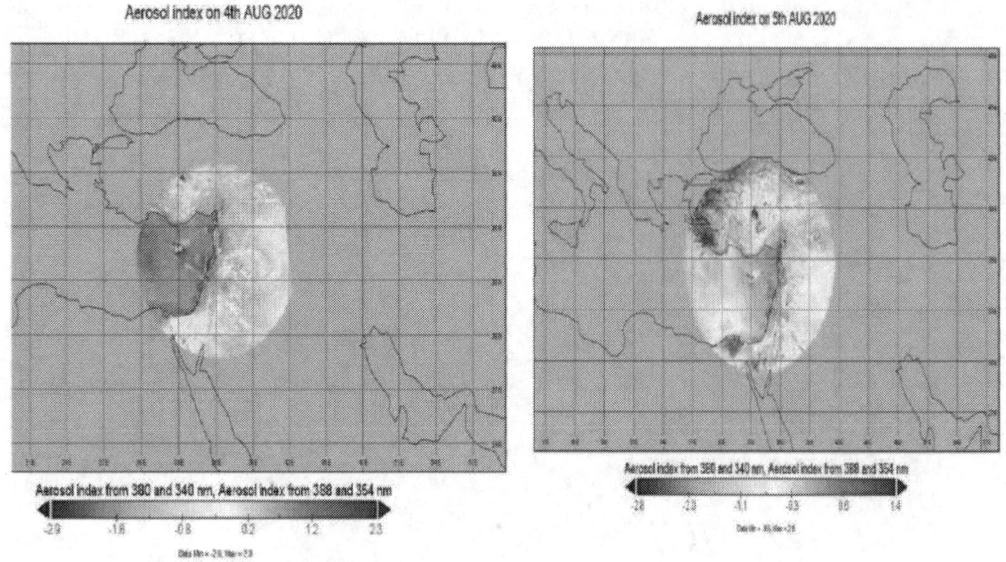

Fig 7 : AOD index map for Beirut city on 4th and 5th Aug 2020

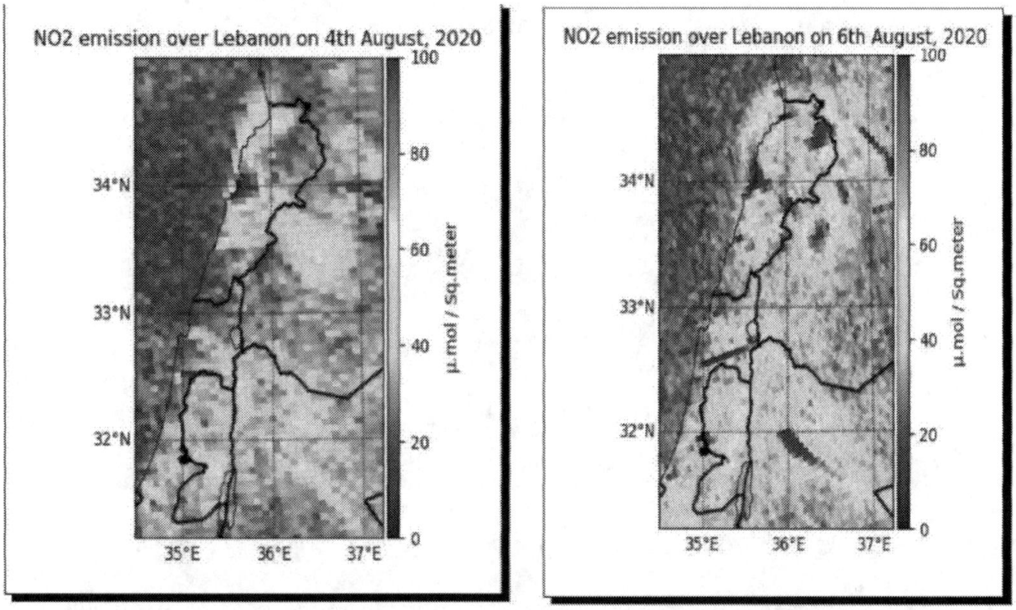

Fig 8 : NO_2 concentration map for Beirut city on 4th and 6th Aug 2020

For damage assessment, a direct comparison of high-resolution Google Earth imagery that was taken just after the explosion over the port of Beirut was carried out. The destroyed buildings can be observed in the image [16]. Both the change detection analysis of Sentinel-1 and Sentinel-2 shows a good correlation with the visual inspection of the high-resolution Google Earth imagery despite the difference in their spatial resolution. Also, throughout the month of August and September, the wind direction is predominantly in the west direction. Therefore, the pollution starts to travel in the South East direction, and the spread of the pollution covered the surrounding countries namely Jordan, Syria, and Israel. The spread covers an area of a 400 km radius.

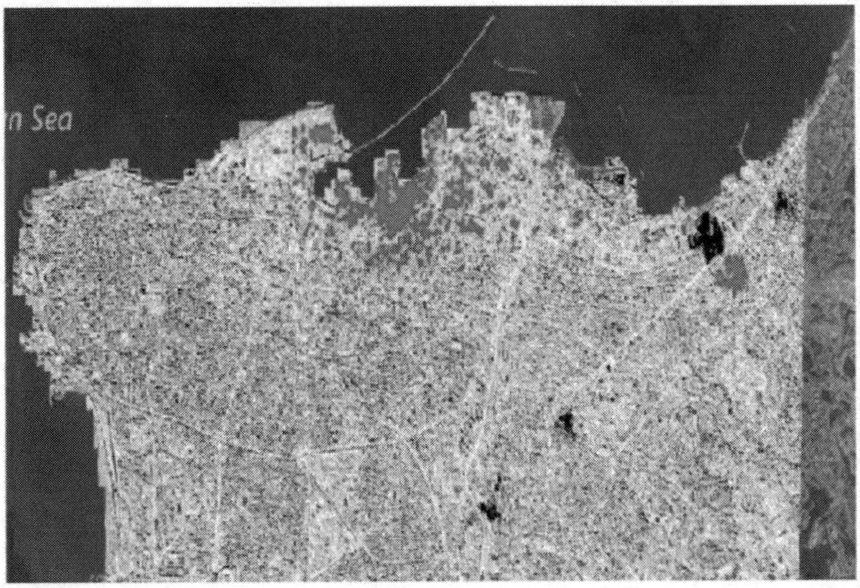

Fig 9 : Sentinel 1 and 2 (Data Assessment)

6. Conclusions

The current study was carried out to find the variation in the aerosol and NO_2 concentration due to the explosion that occurred at the port of Beirut [17]. The massive explosion and regional transport of pollutants between 4th August to 11th November 2020 in Beirut was analyzed by integrating multiple satellite remote sensing products with surface observations.

Aerosol and NO_2 concentrations were monitored regularly and it took nearly a longer period of more than 3 months to get lesser concentration. When ammonium nitrate explodes it releases toxic gases like ammonia and nitrogen dioxide in large amounts when compared to particulate matter. Satellite-derived AOD and NO_2 vertical columns show an excellent correlation between ground-based PM 2.5 and NO_2. A damage assessment was carried out to estimate the damages caused by the explosion in and around the port area. The results obtained from this study can be used as an early warning system and decision support system for mitigating losses due to the explosion of ammonium nitrate.

Cloud cover proved to be a great limitation with the majority of the portions of AOD and NO_2 data unavailable on particular days and hence proved difficult during daily analysis. Only PM2.5 and NO_2 were considered for Air Quality analysis here, however, there are other pollutants such as PM10, CO, SO_2, etc., which can also be considered for further detailed analysis. Another possible development is the integration of satellite observation with ancillary datasets such as wind speed and wind direction. By adding ancillary datasets, it is possible to improve the accuracy of the result.

7. References

[1] B.-A. Wadvalla, "Air pollution threatens health in Beirut," *Nature Middle East*, May 2011, doi: 10.1038/nmiddleeast.2011.60.

[2] A. Siddiqui, S. Halder, P. Chauhan, and P. Kumar, "COVID-19 Pandemic and City-Level Nitrogen Dioxide (NO2) Reduction for Urban Centres of India," *Journal of the Indian Society of Remote Sensing*, vol. 48, no. 7, pp. 999–1006, Jul. 2020, doi: 10.1007/s12524-020-01130-7.

[3] D. J. Miller *et al.*, "Assessing boreal forest fire smoke aerosol impacts on U.S. air quality: A case study using multiple data sets," *Journal of Geophysical Research: Atmospheres*, vol. 116, no. D22, p. n/a-n/a, Nov. 2011, doi: 10.1029/2011jd016170.

[4] K. Schuster, "Dust and garbage loom large over Beirut recovery," *Deutsche Welle*, Aug. 27, 2020. [Online]. Available: https://www.dw.com/en/beirut-pollution-explosion-lebanon-dust-environment-garbage/a-54704363

[5] L. Shikwambana, P. Mhangara, and N. Mbatha, "Trend analysis and first time observations of sulphur dioxide and nitrogen dioxide in South Africa using TROPOMI/Sentinel-5 P data," *International Journal of Applied Earth Observation and Geoinformation*, vol. 91, p. 102130, Sep. 2020, doi: 10.1016/j.jag.2020.102130.

[6] P. Gupta and S. A. Christopher, "Particulate matter air quality assessment using integrated surface, satellite, and meteorological products: Multiple regression approach," *Journal of Geophysical Research*, vol. 114, no. D14, Jul. 2009, doi: 10.1029/2008jd011496.

[7] S. S. Somvanshi, A. Vashisht, U. Chandra, and G. Kaushik, "Delhi Air Pollution Modeling Using Remote Sensing Technique," in *Handbook of Environmental Materials Management*, Cham: Springer International Publishing, 2019, pp. 1–27. Accessed: Jun. 03, 2023. [Online]. Available: http://dx.doi.org/10.1007/978-3-319-58538-3_174-1

[8] Y. Liu, R. A. Kahn, A. Chaloulakou, and P. Koutrakis, "Analysis of the impact of the forest fires in August 2007 on air quality of Athens using multi-sensor aerosol remote sensing data, meteorology, and surface observations," *Atmospheric Environment*, vol. 43, no. 21, pp. 3310–3318, Jul. 2009, doi: 10.1016/j.atmosenv.2009.04.010.

[9] I. Ialongo, H. Virta, H. Eskes, J. Hovila, and J. Douros, "Comparison of TROPOMI/Sentinel-5 Precursor NO & lt; sub>2</sub & gt; observations with ground-based measurements in Helsinki," *Atmospheric Measurement Techniques*, vol. 13, no. 1, pp. 205–218, Jan. 2020, doi: 10.5194/amt-13-205-2020.

[10] Z. Darynova *et al.*, "Evaluation of NO2 column variations over the atmosphere of Kazakhstan using satellite data," *Journal of Applied Remote Sensing*, vol. 12, no. 04, p. 1, Oct. 2018, doi: 10.1117/1.jrs.12.042610.

[11] M. J. Bechle, D. B. Millet, and J. D. Marshall, "Remote sensing of exposure to NO2: Satellite versus ground-based measurement in a large urban area," *Atmospheric Environment*, vol. 69, pp. 345–353, Apr. 2013, doi: 10.1016/j.atmosenv.2012.11.046.

[12] C.-A. Lin, Y.-C. Chen, C.-Y. Liu, W.-T. Chen, J. H. Seinfeld, and C. C.-K. Chou, "Satellite-Derived Correlation of SO2, NO2, and Aerosol Optical Depth with Meteorological Conditions over East Asia from 2005 to 2015," *Remote Sensing*, vol. 11, no. 15, p. 1738, Jul.

2019, doi: 10.3390/rs11151738.

[13] A. P. Sati and M. Mohan, "Analysis of air pollution during a severe smog episode of November 2012 and the Diwali Festival over Delhi, India," *International Journal of Remote Sensing*, vol. 35, no. 19, pp. 6940–6954, Sep. 2014, doi: 10.1080/01431161.2014.960618.

[14] C. Mozumder, K. V. Reddy, and D. Pratap, "Air Pollution Modeling from Remotely Sensed Data Using Regression Techniques," *Journal of the Indian Society of Remote Sensing*, vol. 41, no. 2, pp. 269–277, Oct. 2012, doi: 10.1007/s12524-012-0235-2.

[15] H. Irie *et al.*, "Validation of OMI tropospheric NO₂ column data using MAX-DOAS measurements deep inside the North China Plain in June 2006: Mount Tai Experiment 2006," *Atmospheric Chemistry and Physics*, vol. 8, no. 22, pp. 6577–6586, Nov. 2008, doi: 10.5194/acp-8-6577-2008.

[16] S. D. Superczynski and S. A. Christopher, "Exploring Land Use and Land Cover Effects on Air Quality in Central Alabama Using GIS and Remote Sensing," *Remote Sensing*, vol. 3, no. 12, pp. 2552–2567, Nov. 2011, doi: 10.3390/rs3122552.

[17] A. Waked *et al.*, "Modeling air pollution in Lebanon: evaluation at a suburban site in Beirut during summer," *Atmospheric Chemistry and Physics*, vol. 13, no. 12, pp. 5873–5886, Jun. 2013, doi: 10.5194/acp-13-5873-2013.

International Conference on Sustainable Practices and Innovations in Civil Engineering (SPICE 2023)
A Survey Study on Challenges and Factors Affecting in Adopting Sustainable Construction Methods in Indian Construction Industry

Uma Ravi Teja Macherla[1]
[1]*Assistant Professor, Department of Civil Engineering, Sri Venkateswara College of Engineering and Technology, Srikakulam, Andhra Pradesh, India-532410;*
Email: *ravimacherla007@gmail.com*

Abstract

Due to India's growing urbanization, the building industry is experiencing severe environmental challenges. Demand for residential housing is rising, which causes people to use more energy, resources, and raw materials leading to the rise of the carbon footprint. The environment is significantly impacted by the construction sector. It uses up natural resources and energy reserves, and it contributes significantly to the world's CO2 emissions. The production of waste, the use of energy and water inefficiently, as well as several other environmental harms, have all been linked to the building sector. The step towards sustainability is getting a good cause due to the extinction of resources and increase of pollution in the environment. As per the sources, the majority of the carbon footprint is produced from cement and cement-related products (Construction Industry). Green Building practices are showing a good impact in the reduction of carbon footprint in the environment from the construction industry. This study aims to explore the importance of sustainable construction by conducting two surveys, a primary survey and a secondary survey. The primary survey focused on identifying the challenges in adopting sustainable construction practices, while the secondary survey aimed to identify the factors that influence the adoption of sustainable construction practices. The research employed the Relative Importance Index (RII) to identify the major challenges faced in adopting sustainable construction practices. With the help of RII factor, I was easily figured out the main challenges are Limited Awareness and Expensive Products.

Keywords: Sustainable Construction, Challenges, Factors, Questionnaire Survey, Carbon Foot Print

1. Introduction

Civil engineering is a branch of engineering that deals with the design, construction, and maintenance of the built environment, including structures such as buildings, bridges, roads, dams, and airports. It also includes the design and management of infrastructure systems, such as water supply and sewage treatment. Civil engineers use principles of physics, mathematics, and materials science to plan, design, and oversee construction projects. They are involved in every stage of a project, from initial concept and feasibility studies to final construction and maintenance. Civil engineers play a crucial role in ensuring that our built environment is safe, efficient, and sustainable. Sustainability in construction refers to the practice of designing, building, and maintaining structures and infrastructure in an environmentally responsible and resource-efficient manner. This includes using materials and methods that minimize negative impacts on the environment and promoting the use of renewable energy sources. It also involves designing buildings and infrastructure that are energy-efficient, reducing the use of non-renewable resources, and promoting the use of sustainable transportation options. Additionally, sustainable construction practices aim to minimize waste, reduce pollution and create healthy, comfortable, and safe indoor environments. Overall, sustainability in construction is about creating a built environment that supports the well-being of current and future generations while preserving natural resources and biodiversity.

© The Authors, published by EDP Sciences. This is an open access article distributed under the terms of the Creative Commons Attribution License 4.0 (https://creativecommons.org/licenses/by/4.0/).

Sustainable construction is important for several reasons. Firstly, it helps to reduce the environmental impact of buildings and infrastructure by minimizing the use of non-renewable resources, reducing waste and pollution, and promoting the use of renewable energy sources. This is important because the built environment is responsible for a significant portion of global greenhouse gas emissions, and sustainable construction practices can help to mitigate this impact. Secondly, sustainable construction helps to create buildings and infrastructure that are more energy-efficient and cost-effective to operate. This is beneficial for both the environment and for building owners and occupants, as it can lead to significant cost savings on energy bills. Thirdly, sustainable construction can help to create healthy, comfortable, and safe indoor environments for building occupants. This is important for promoting well-being and productivity, and can also help to attract and retain tenants or residents. Lastly, sustainable construction can also help to promote social sustainability by creating spaces that are accessible to all members of society, and by supporting local economies through the use of locally-sourced materials and labor.

Overall, sustainable construction is important for addressing the pressing environmental challenges of our time, while also providing economic and social benefits. It is a crucial step towards creating a more sustainable built environment that supports the well-being of current and future generations. Awareness alone may not be enough to significantly increase the sustainable construction footprint. While raising awareness about the benefits of sustainable construction and the environmental impact of the built environment can help to encourage more sustainable practices, it is not sufficient on its own. To truly increase the sustainable construction footprint, some other factors also need to be considered and addressed. For example, a lack of regulations and standards for sustainable construction can make it difficult for builders and developers to implement sustainable practices. Additionally, the availability and cost of sustainable materials and technologies can also be a barrier to the wider adoption of sustainable construction. Furthermore, sustainable construction practices can often require a higher initial investment, which can be a barrier for some developers and building owners. To overcome these barriers, incentives, and regulations are required to promote sustainable construction, and to make it more accessible and cost-effective.

1.1 Objectives

- To identify the current level of awareness and knowledge among construction industry professionals regarding sustainable practices
- To identify the barriers and challenges faced by construction professionals adopting sustainable practices
- To propose recommendations for overcoming the barriers and challenges.
- To identify the key factors in adopting sustainable practices

2. Literature Review

Ahmed Abu Hanieh et.al the purpose of the study is to create a sustainable model for the Palestinian residential construction industry by following the industry from conception to completion. The sustainability model will specify each resource input and link it to the outputs of the construction process, which include all dynamic contributors—labor, water, and energy. For a 100 m2 residential construction, a model is implemented based on the sustainable supply chain. All relevant expenses have been included to indicate the building's overall cost for the implemented case study. The Value Creation Index for the case under investigation has been estimated and displays the relationship between the profits made by neighborhood construction firms and the costs associated with the building. The substantial resource use highlights the significance and necessity of developing a sustainability model for the industry.

Prithviraj Dilip Mane stated that Sustainable building approaches have become more and more popular in India. Newspaper headlines frequently discuss the growing concern over the negative consequences of construction-related activities and the need to remedy them. For improved environmental and social protection, the Indian government has already taken significant steps to promote the idea of green building. Due to growing environmental concerns, developers should review their current construction methods and implement sustainable construction techniques in their upcoming projects. The degree of awareness, knowledge, and comprehension of the effects of each activity, however, determines how quickly all the connected developers and builders adopt this part of the construction. To measure the differences in knowledge, awareness, and application of sustainable practices based on the perspectives of the project developers in India, a survey has been organized in the current study that is in line with this purpose. The right steps should be taken to expand this understanding at all development levels to increase the momentum and acceptability of sustainable behavior in the sector.

Kaanchan M. Patil et.al Due to India's growing urbanization, the building industry is experiencing severe environmental issues. Demand for residential housing is rising, which causes people to use more energy, resources, and raw materials, increasing their carbon footprint. Environmental impact problems, such as altered weather patterns and ecological damage, already exist in all urban areas. The use of sustainable technology and materials in development is the answer. Sustainable development satisfies current needs without sacrificing the capacity of future generations to satisfy their own needs. This essay elaborates on the idea of sustainable development, which has gained prominence over the past 20 years. The utilization of environmentally friendly building materials and technologies will enable this progress. The idea acknowledges that nature and human civilization go hand in hand and that for the human community to continue, nature must be conserved and perpetuated. This concept is expressed through designs that demonstrate conservation principles, and sustainable development urges us to apply such principles to our daily lives. A different strategy than the conventional design must be used for sustainable development. The new strategy must take into account how each material and the technological decision will affect local, regional, and global natural and cultural resources.

A O Oluwunmi et. al order to create a sustainable environment, development must embrace the green construction concepts. However, there is relatively little research on its use in educational institutions, and the few studies that are accessible in Nigeria do not prioritize students' viewpoints. Therefore, this study looked into how Covenant University students in Ota, Ogun State, Nigeria, perceived the necessity to adopt green building principles. A response rate of 94% was attained out of 101 surveys distributed to students in the university's departments of architecture, building technology, and estate management. Tables were used to exhibit the data after it had been analyzed using descriptive statistics like percentage, weighted mean, and relative importance index (RII). According to the report, 88.42% of the students are familiar with the idea of green construction. The level of adoption of green building standards at Covenant University is quite low, according to the students. Additionally, according to the students, the university may profit from enhanced indoor air and water quality, less pollution and environmental deterioration, energy efficiency, and water conservation by implementing green building concepts.

Vanderley M. John stated Engineers face a significant opportunity and challenge in the form of sustainable development, which calls for a radical rethinking of the entire industrial sector to reduce the environmental impacts of products' life cycles by a factor of five while simultaneously boosting their economic and social benefits. Without fundamental changes to business practices and the acceptance of innovations across all industrial sectors, this cannot be accomplished [Crul & Diehl 2007; Eppinger & Hopkins 2010]. To achieve global sustainability, the building industry is viewed as a priority. The industry consumes between 40 and 75 percent of the total value of materials harvested and contributes up to 15% of the GDP, making it the sector with the highest intensity of material used in the overall economy. The amount of waste produced along the supply chain is inversely correlated with the overall cost of the resources used. The life cycle assessment of construction items, which mostly consist of buildings and roads, revealed that they account for a sizeable portion of total energy use. About 25% of all CO_2 gas emissions from the production of energy are caused by buildings themselves when they are in the "in-use" phase [Price et al. 2006]. Cement manufacturing accounts for 5% of all anthropogenic carbon emissions.

3. Methodology

A detailed analysis has been carried out, and the results are described in the following Flowchart

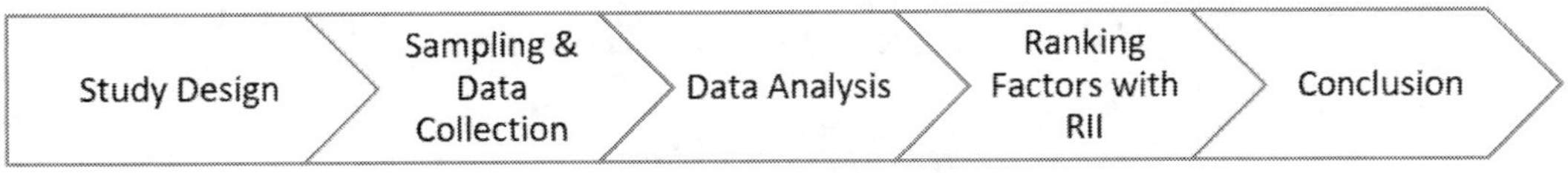

Fig A Methodology Chart

- Study Design

  Study design refers to the plan or strategy used to conduct research and collect data for analysis.
- Sampling & Data Collection

  Data collection involves gathering information through various methods, such as surveys, interviews, observations, or experiments, for analysis and interpretation.
- Data Analysis

  Data analysis is the process of examining and interpreting data to draw conclusions, identify patterns, and test hypotheses.
- Ranking Factors with RII

  Ranking factors using RII (Relative Importance Index) involves assigning weights to factors based on their perceived importance in achieving a goal.
- Conclusions

4. Results and Discussions

In this Survey, We mostly encouraged all the participants to know about the importance of Sustainable Construction and allowed them to freely express their interest in Sustainable Construction. All the respondents have a little bit of knowledge of it but mainly all the respondents are saying that this type of structure will be accepted by the High Rise construction people but it will not be accepted by Middle-Class Housing areas due to their initial investment. A few of the Respondents suggested that it should be implemented in the education system to encourage and create awareness of sustainable construction that how it saves the environment even though its initial investment is high but the life of the structure and eco-friendly environment will be a key factor over it. In this survey, I found students showed interest to learn a new concept in the construction industry and they have shown very much interest to know what products will come under Green products or Eco-Friendly Products. This survey data sheet Consists of Respondent Profiles, Challenges to Adopt Sustainable Construction, and Factors Affecting the Acceptance of Sustainable Construction

4.1 Respondent Profile

Respondent profile refers to the characteristics of individuals who participate in a study or survey, such as age, gender, education, income, and other relevant demographic or behavioral factors.

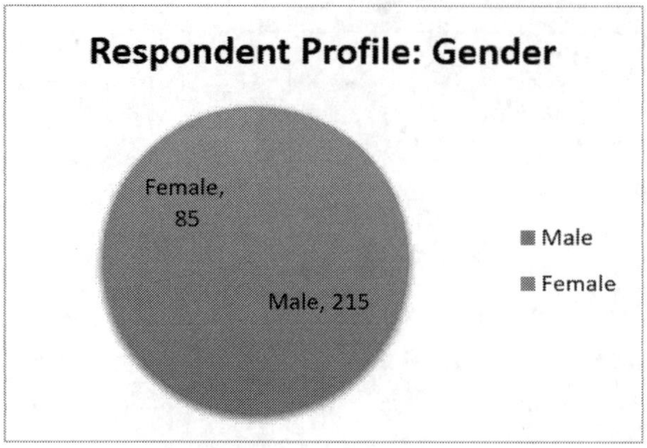

Fig 1 Respondent Profile- Gender

In this survey, the respondent profile was predominantly male, with 215 male participants and 85 female participants.

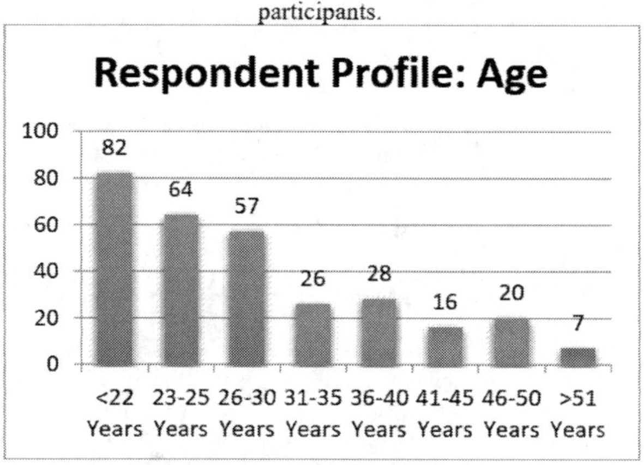

Fig 2 Respondent Profile- Age

The respondent profile in terms of age was diverse, with the majority falling between 23-25 years (64 participants) and <22 years (82 participants), while the age range of 51 years had the lowest participation with only 7 participants.

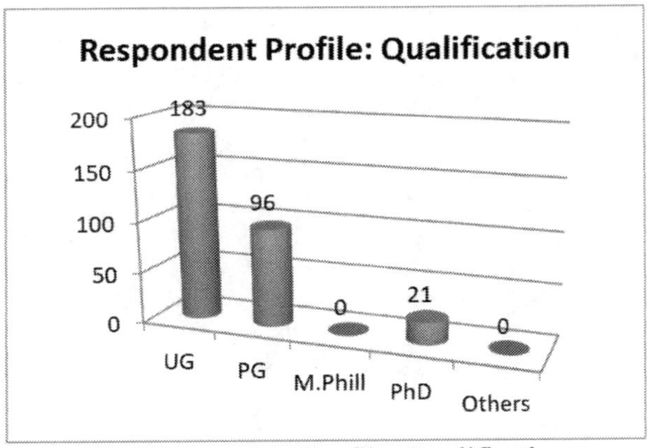

Fig 3 Respondent Profile- Qualification

The educational qualification of the respondents varied, with the majority having an undergraduate degree (183 participants), followed by postgraduate degrees (96 participants), while no respondents had a Master of Philosophy degree and only 21 had a Ph.D., with none falling under the "Others" category.

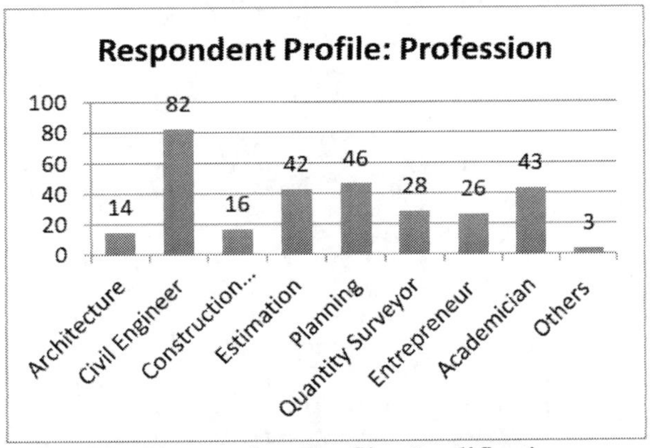

Fig 4 Respondent Profile- Qualification

The profession of the respondents was diverse, with civil engineers being the most common (82 participants), followed by architects (14 participants) and entrepreneurs (26 participants). Other professions, such as construction project managers, estimators, planners, quantity surveyors, academicians, and individuals in other fields, were also represented in the respondent profile.

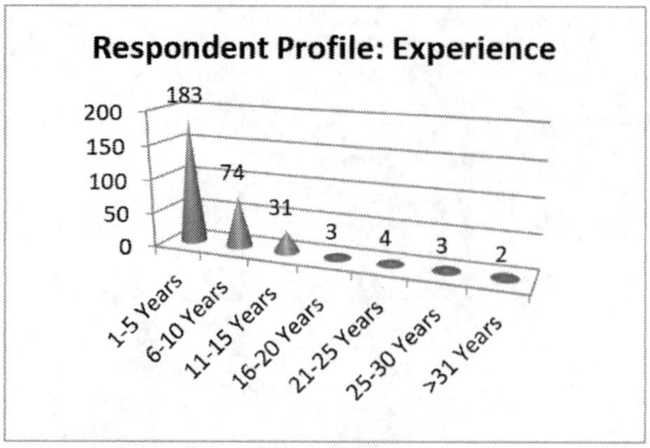

Fig 5 Respondent Profile- Experience

The experience of the respondents in the construction industry varied, with the majority having 1-5 years of experience (183 participants), followed by 6-10 years of experience (74 participants). Only a small number of respondents had more than 15 years of experience, with only 2 having more than 31 years of experience.

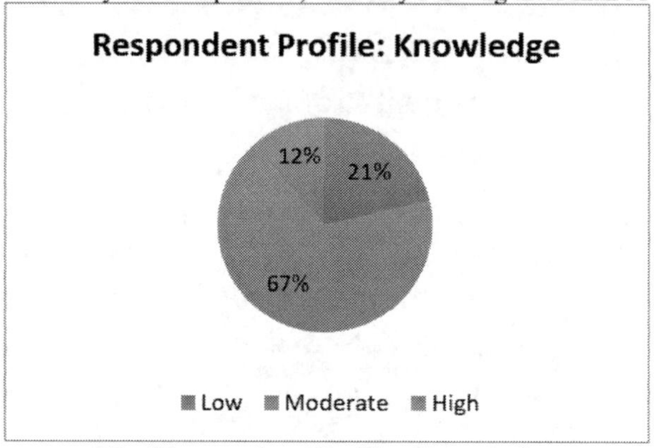

Fig 6 Respondent Profile- Knowledge

The knowledge of respondents on green buildings varied, with the majority having a moderate level of knowledge (201 participants), followed by low knowledge (64 participants) and high knowledge (35 participants).

4.2 Challenges in Adopting Sustainable Construction

 https://doi.org/10.1051/matecconf/202338401004

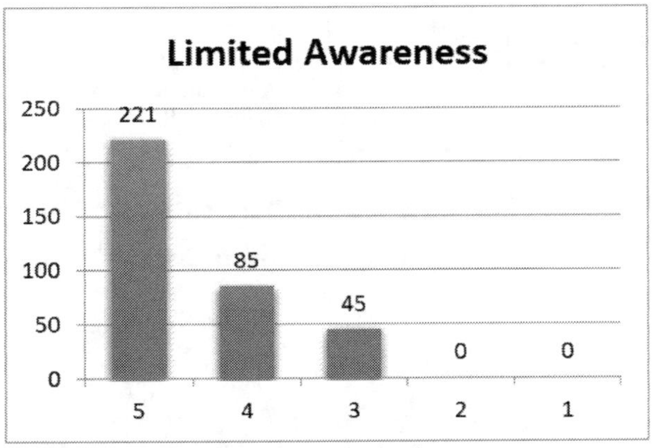

Fig 7 Challeneges- Limited Awareness

In regards to the limited awareness challenge, the majority of respondents strongly agreed that it was a challenge (221 participants), followed by those who agreed (34 participants). No respondents were neutral, disagreed, or strongly disagreed with this challenge.

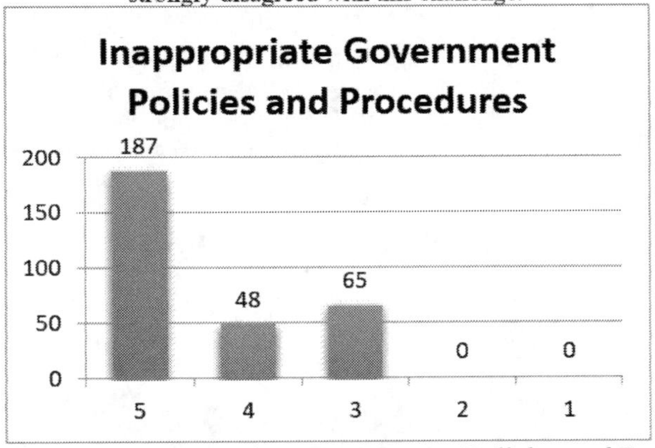

Fig 8 Challenges- Inappropriate Government Policies and Procedures

In regards to the inappropriate government policies and procedures challenge, the majority of respondents strongly agreed that it was a challenge (187 participants), followed by those who agreed (48 participants). A significant number of respondents were neutral about this challenge (65 participants), while none of the respondents disagreed or strongly disagreed with it.

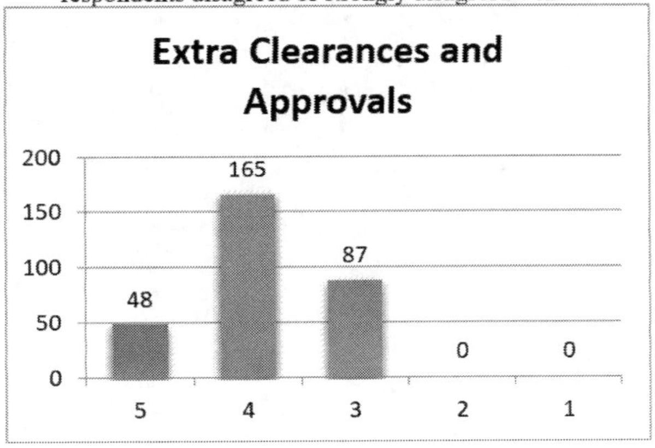

Fig 9 Challeneges- Extra Clearances and Approvals

In regards to the extra clearances and approvals challenge, the majority of respondents agreed that it was a challenge (165 participants), followed by those who were neutral about it (87 participants). A smaller number of respondents strongly agreed that it was a challenge (48 participants), while none of the respondents disagreed or strongly disagreed with this challenge.

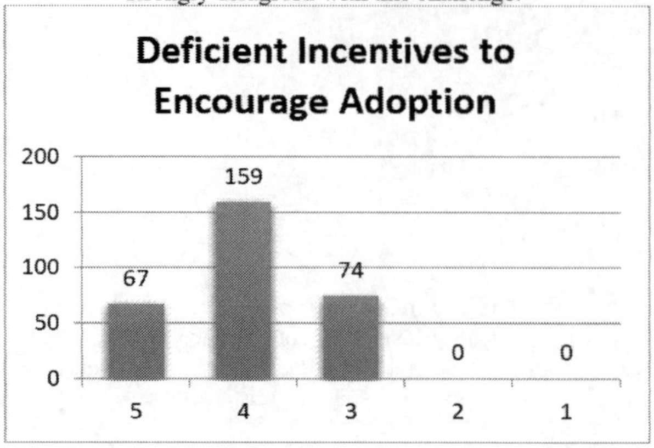

Fig 10 Challeneges- Deficient Incentives to Encourage Adoption

In regards to the deficient incentives to encourage adoption challenge, the majority of respondents agreed that it was a challenge (159 participants), followed by those who were neutral about it (74 participants). A smaller number of respondents strongly agreed that it was a challenge (67 participants), while none of the respondents disagreed or strongly disagreed with this challenge.

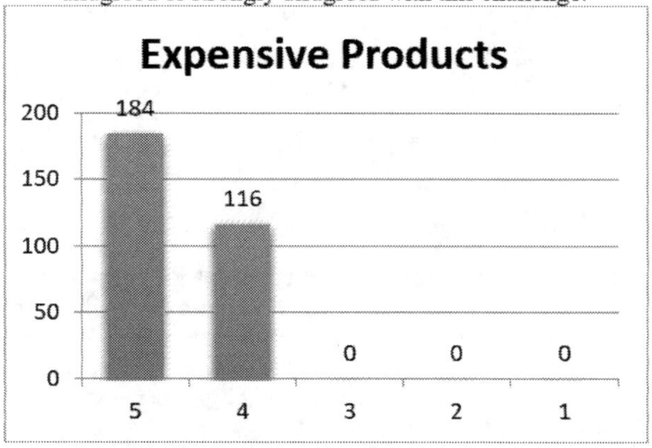

Fig 11 Challeneges- Expensive Products

In regards to the expensive products challenge, the majority of respondents strongly agreed that it was a challenge (184 participants), followed by those who agreed (116 participants). None of the respondents were neutral, disagreed, or strongly disagreed with this challenge.

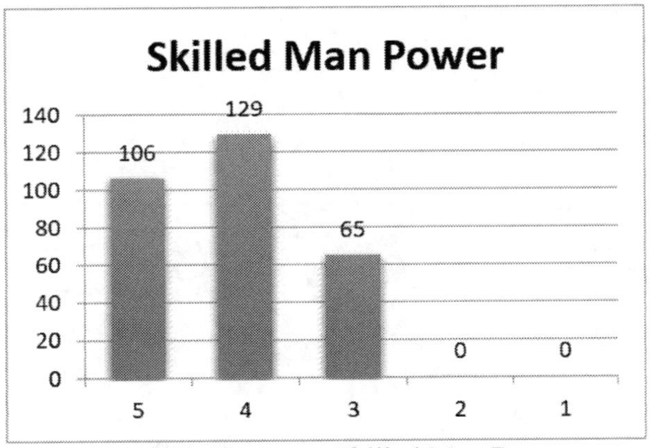

Fig 12 Challeneges- Skilled Man Power

In regards to the skilled manpower challenge, the majority of respondents agreed that it was a challenge (129 participants), followed by those who strongly agreed (106 participants). A significant number of respondents were neutral about this challenge (65 participants), while none of the respondents disagreed or strongly disagreed with it.

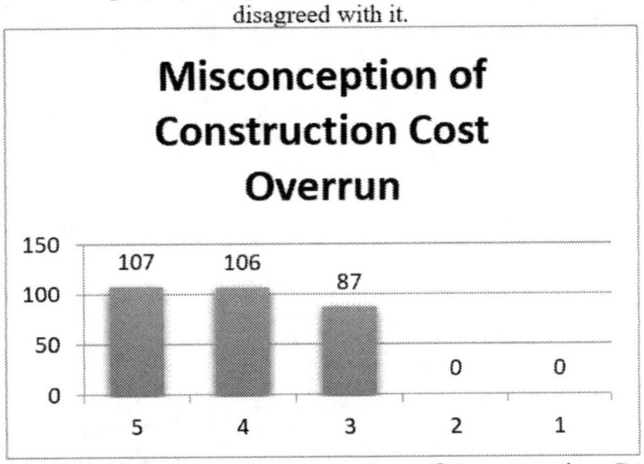

Fig 13 Challeneges- Misconception of Construction Cost

In regards to the misconception of construction cost overrun challenge, the majority of respondents either strongly agreed or agreed that it was a challenge (107 and 106 participants, respectively). A significant number of respondents were neutral about this challenge (87 participants), while none of the respondents disagreed or strongly disagreed with it.

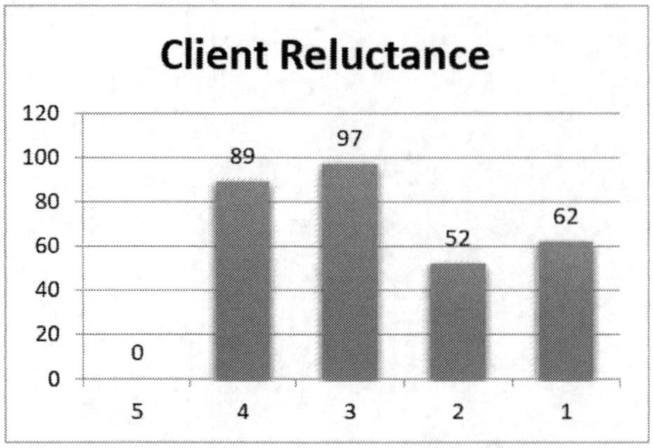

Fig 14 Challeneges- Client Reluctance

In regards to the client reluctance challenge, the majority of respondents were either neutral (97 participants) or disagreed (52 participants) that it was a challenge. A significant number of respondents agreed that it was a challenge (89 participants), while none of the respondents strongly agreed with it. However, a considerable number of respondents also strongly disagreed with this challenge (62 participants).

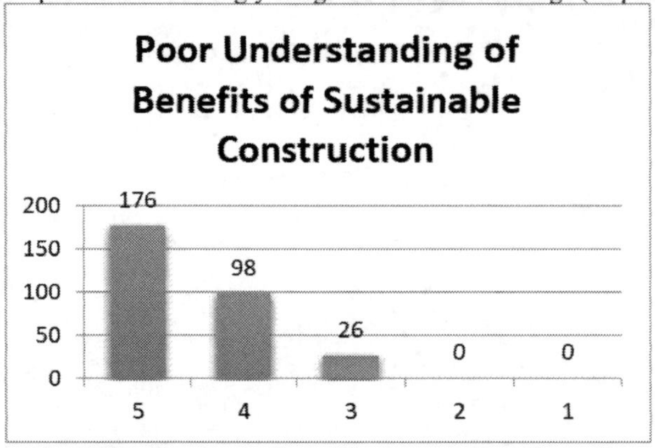

Fig 15 Challeneges- Poor Understanding Benefits

According to the survey, a majority of the respondents agreed that the challenge of poor understanding of benefits of sustainable construction exists, with 176 participants strongly agreeing and 98 participants agreeing with this challenge. A small number of respondents were neutral towards this challenge (26 participants), while none of the respondents disagreed or strongly disagreed with it.

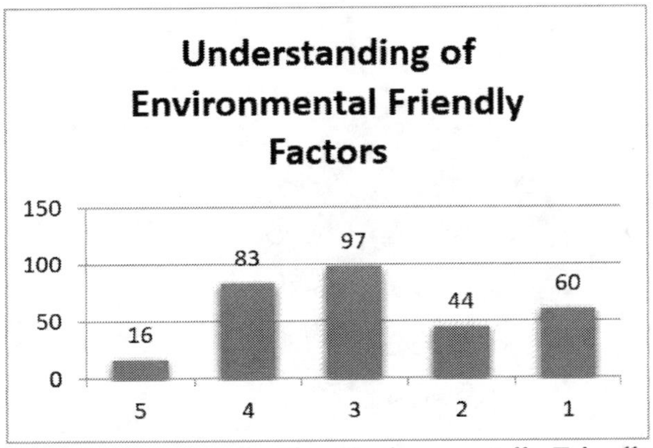

Fig 16 Challeneges- Understanding Environmentally Friendly Factors
According to the survey results, for the challenge of Understanding of Environmental Friendly Factors, 16 respondents strongly agreed, 83 agreed, 97 were neutral, 44 disagreed, and 60 strongly disagreed.

4.3 Factors Affecting Acceptance of Sustainable Construction

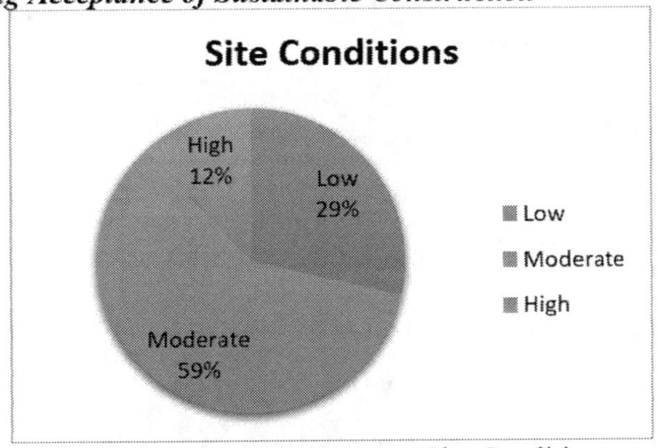

Fig 17 Environmental Factors-Site Conditions
In the Environmental Factors, Respondents stated that Site Conditions will be the moderate factor in Accepting Sustainable Construction.

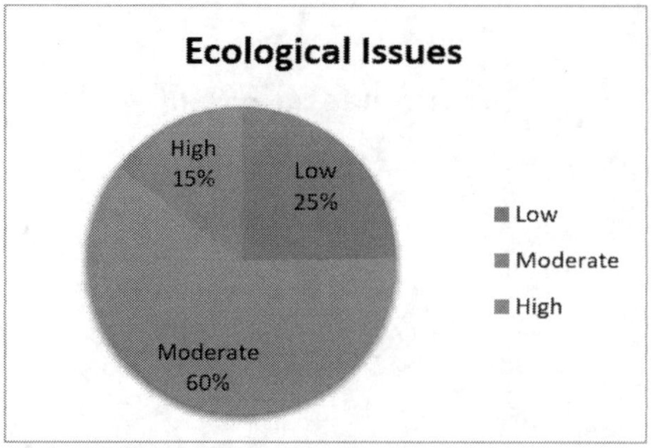

Fig 18 Environmental Factors-Ecological Issues

In the Environmental Factors, Respondents stated that Ecological Issues will be the moderate factor in Accepting Sustainable Construction.

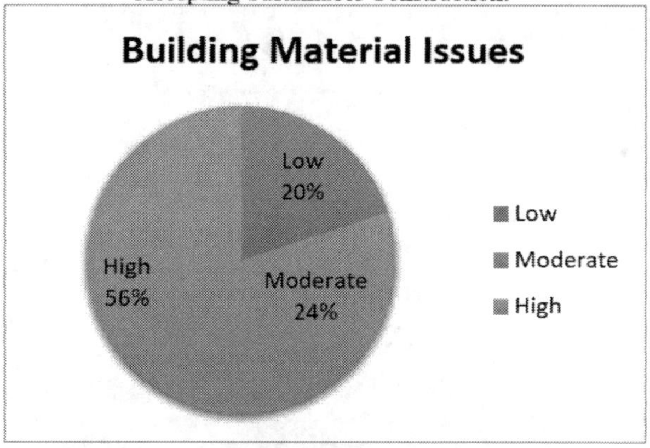

Fig 19 Environmental Factors-Building Material Issues

In the Environmental Factors, Respondents stated that Building Material Issues will be the High factor in Accepting Sustainable Construction.

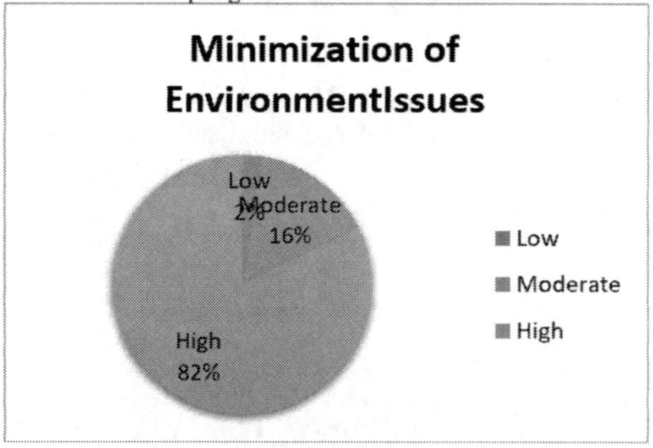

Fig 20 Environmental Factors-Minimization of Environment Issues

In the Environmental Factors, Respondents stated that Minimization of Environment Issues will be the High factor in Accepting Sustainable Construction.

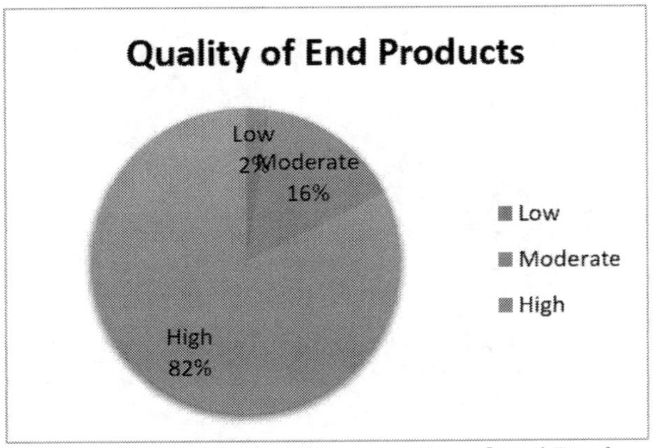

Fig 21 Environmental Factors-Quality of End Products
In the Environmental Factors, Respondents stated that the Quality of the End Product will be the High factor in Accepting Sustainable Construction.

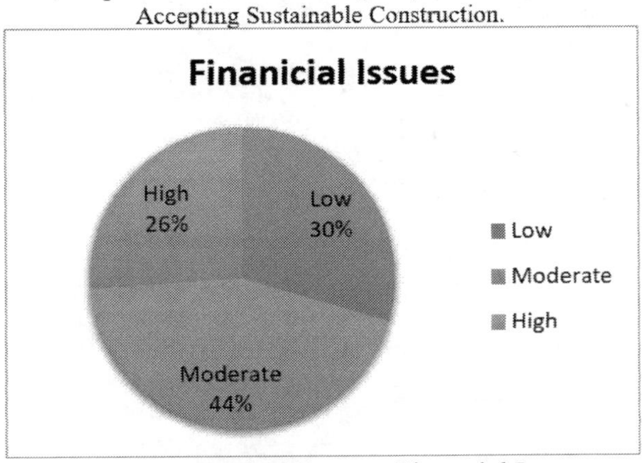

Fig 22 Economical Factors-Financial Issues
In the Economical Factor, Respondents stated that Financial Issues will be the Moderate Factor for Accepting Sustainable Construction.

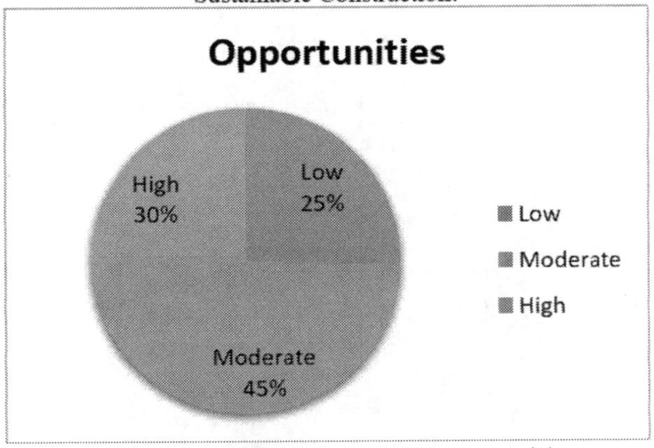

Fig 23 Economical Factors-Opportunities
In the Economical Factor, Respondents stated that Opportunities will be the Moderate Factor for Accepting Sustainable Construction.

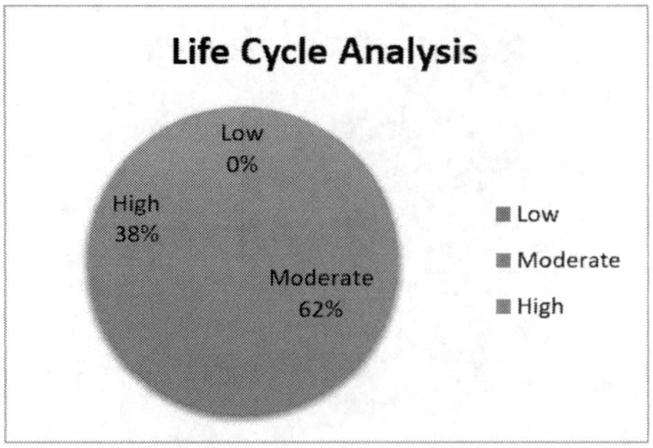

Fig 24 Economical Factors-Life Cycle Analysis

In the Economical Factor, Respondents stated that Life Cycle Analysis will be the Moderate Factor for Accepting Sustainable Construction.

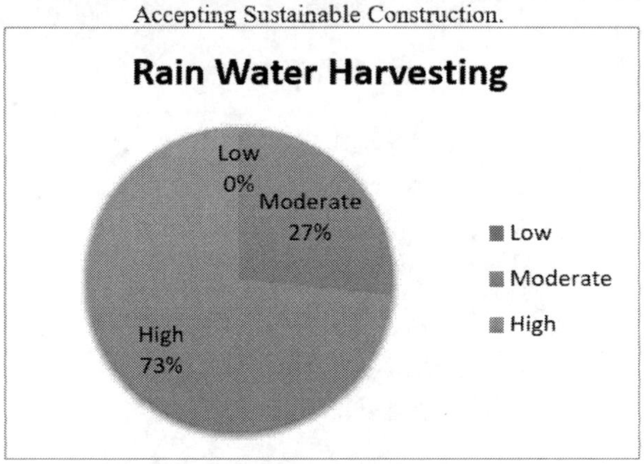

Fig 25 Technological Factors-Rain Water Harvesting

In the Technological Factor, Respondents stated that Rain Water Harvesting Techniques will be the High Factor for Accepting sustainable Construction.

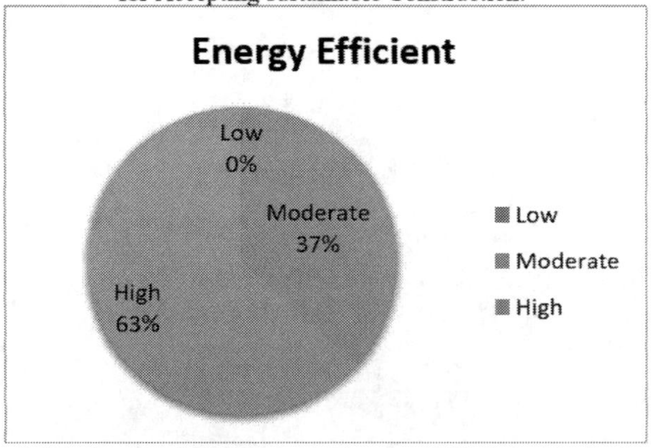

Fig 26 Technological Factors-Energy Efficient Techniques

In the Technological Factor, Respondents stated that Energy Efficient Techniques will be the High Factor for Accepting sustainable Construction.

https://doi.org/10.1051/matecconf/202338401004

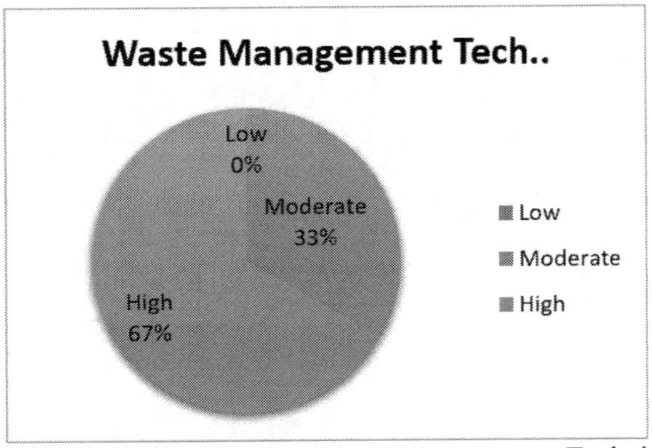

Fig 27 Technological Factors-Waste Management Techniques

In the Technological Factor, Respondents stated that Waste Management Techniques will be the High Factor for Accepting sustainable Construction.

4.4 Relative Importance Index(RII):

The Relative Importance Index (RII) is a statistical tool used to measure the importance of various factors in a given study or survey. It is often used in market research and other social sciences to identify the most significant variables or factors that influence a particular outcome.

S.no	Challenges	RII FACTOR	MEAN
1	Limited Awareness	0.92	4.587
2	Inappropriate Government Policies and Procedures	0.88	4.407
3	Extra Clearances and Approvals	0.77	3.870
4	Deficient Incentives to Encourage Adoption	0.80	3.977
5	Expensive Products	0.92	4.613
6	Skilled Man Power	0.83	4.137
7	Misconception of Construction Cost Overrun	0.81	4.067
8	Client Reluctance	0.54	2.710
9	Poor Understanding of Benefits of Sustainable Construction	0.90	4.500
10	Understanding of Environmental Friendly Factors	0.57	2.837

5. Conclusions

The following conclusions are from this study:

1. Due to Less number of people available in our field, I'm unable to reach more people.
2. All the respondents are a little bit aware of Sustainable Construction.
3. In this survey, I found the students aren't aware of this concept. Mainly they should study the concepts of Sustainable Construction or at least they should participate in conferences or project works.
4. Experienced people are good at this type of construction and they are very much interested to adopt this type of Eco-Friendly Construction.
5. In the above study, all the respondents were very much interested to adopt technological factors in their constructions.
6. This survey mainly worked on the concept of finding out the challenges in adopting sustainable constructions and mainly awareness is required.

7. The Cost factor for the products and workmanship plays a major role in adopting sustainable construction
8. Few people n the respondents are not aware of Life Cycle Analysis/ Assessment.
9. Building Materials issues will have a major effect on adopting Sustainable construction
10. Client Approval also is one of the major challenges in adopting sustainable constructions.
11. Respondents Stated that Limited Awareness and Expensive Products are the Main Reason to adopt Sustainable Construction. They have RII factor with 0.92.

References

Journal article

G AlSanad, Shaikha. "Awareness, Drivers, Actions, and Barriers of Sustainable Construction in Kuwait." *International Conference on Sustainable Design, Engineering and Construction*, 2015: 969-983.

Ar.Kaanchan M. Patil, and Er.Mahendraa S. Patil. "Sustainable Construction Materials & Technology in Context with Sustainable Development." *International Journal of Engineering Research and Technology*, 2017: 112-117.

Clinton Aigbavboaa, Ifije Ohiomaha, and Thulisile Zwanea. "Sustainable construction practices: "a lazy view" of construction professionals in the South Africa construction industry." *Energy Procedia*, 2017: 3003-3010.

David William Dobson1, Amr Sourani, Begum Sertyesilisik, and Ashley Tunstall. "Sustainable Construction: Analysis of Its Costs and Benefits." *American Journal of Civil Engineering and Architecture*, 2013: 32-38.

Hanieh, Ahmed Abu, Mohammad Karaeen, and Afif Akel Hasan. "Sustainability Model from Cradle-to-Gate: Residential Buildings in Palestine." *International Journal of Sustainable Construction Engineering and Technology*, 2022: 1-14.

Jamilus Md Hussin, Ismail Abdul Rahman, and Aftab Hameed Memon. "The Way Forward in Sustainable Construction: Issues and Challenges." *International Journal of Advances in Applied Sciences*, 2013: 15-24.

Mr. Prithviraj Dilip Mane. "Green Buildings and Sustainable Construction." *International Journal of Engineering Research & Technology*, 2017: 362-376.

O Oluwunmi O P Oladayo, A Role, and T O Afolabi. "Benefits and Barriers to the Implementation of Green Building Standards in Universities: What are Students' Views?Standards in Universities: What are Students' Views?" *International Conference on Sustainable Infrastructural Development*, 2019: 1-11.

Olanipekun, Ayokunle Olubunmi, Xia Bo, and Skitmore Martin. *Green Building Incentives: A review.* Brisbane Australia: Queensland University of Technology, , 2016.

Vanderley M. John. "Sustainable Construction, Innovation and Durability: Trends." *International Conference on Durability of Building Materials and Components*, 2015: 1-7.

Development of alternatives for Signalized intersection- A Case study of Laxmi Mill junction, Coimbatore

P. Indhiradevi[1], *B.S.Meenakshi*[2], *M. Nazir Ahamed*[3], *M. Ram Vivekanandhan*[4], *B. Aravind madhavan*[5], *D. Santhosh Kumar*[6]

[1,2&3]Assistant Professor, Department of Civil Engineering, KPR Institute of Engineering and Technology, Coimbatore. Email id: Indhiradevi.p@kpriet.ac.in
[4]PG Student, Department of Civil Engineering, KPR Institute of Engineering and Technology, Coimbatore.
[5,6]Under Graduate Student, Department of Civil Engineering, KPR Institute of Engineering and Technology, Coimbatore.

Abstract. Urbanization and industrialization are two of the most important features of modern civilization. This leads to traffic congestion at the major roads of urbanized areas. At intersections, particularly those where four roads converge, signalization is also regarded as a more challenging issue. It takes longer for cars to cross that intersection and move from one side of the roads to the other, especially during rush hour. This may lead to heavy traffic congestion at the junction and may cause accidents. This paper presents the proposed alternatives for signalized intersection at Laxmi mill junction, Coimbatore. To control the flow of heavy traffic and also make sure the vehicles are moving continuously without waiting at the intersection, following two alternatives have been proposed. a) To design the medians instead of the signals with pedestrian crossing bridge in junction b) To design the roundabout at the intersection with pedestrian bridge. By providing these alternatives, Junction traffic flow rates can be managed and accident rate also will get reduced in that junction

Keywords: Intersection, Pedestrian crossing, Peak hours, Roundabout, Signalization, Urbanization

1 Introduction

Over the past few decades, the majority of cities have experienced a hazardous growth of industrialization and urbanisation, and Coimbatore is no different. In urbanised and industrialised areas, especially during peak hours, traffic congestion is growing every day as a result of the growing population. This causes delays, aggravation, and environmental pollution. Therefore, controlling all of the quiet automobiles manually at the intersection is exceedingly difficult. Despite the presence of signals at both intersections, the level of traffic congestion and time delays has not been adequately decreased.

© The Authors, published by EDP Sciences. This is an open access article distributed under the terms of the Creative Commons Attribution License 4.0 (https://creativecommons.org/licenses/by/4.0/).

Due to the city's rapid growth and the construction of colleges, hospitals, an international airport, and other facilities, the number of traffic conflicts near Laxmi Mill Junction has increased. Therefore, various alternatives (unsignalized crossings) are offered and created in accordance with IRC regulations in order to address the issues with signalised intersections. Unsignalized junctions are typically created by building medians where necessary to control traffic flow or by adding a roundabout in the middle. If these options are available at the crossroads, then both traffic congestion and the amount of time that vehicles must wait at the intersection will be significantly decreased.

2 Scope & Objective of the Project:

The project's primary goal is to offer alternatives for signalised intersections, particularly for four lane intersections. The following two alternatives have been proposed instead of signals to control the traffic flow.

a) To provide the medians instead of signals at intersection

b) To design the roundabout at the intersection.

In this work, we examined the movement of vehicles at the main intersection and intended to eliminate the intersection's traffic signals and build a median or roundabout to regulate traffic.

3 Literature Review

Literature study is done for both median construction and roundabout construction which helps to make the project more effective.
Frank Gross et al (1) estimated the security effects of modifying signal-controlled intersections to roundabouts. During this paper the benefits of modifying the signalized intersection into roundabout is highlighted. (i) Central islands have lesser conflict points when compared to signal-controlled intersections. (ii) Crash frequency is reduced when signal-controlled intersections are converted to roundabouts. (iii) Roundabouts at signal-controlled crossings reduce the severity of crashes. (iv) Safety effect findings may differ by volume count of traffic, location, and approaches count and number of road lanes.
All-way stop-controlled (AWSC) junctions and rotary intersections were compared using the aaSIDRA software tool by Evdokia Vlahos et al. (2) after calibrating it using a set of reliable local conditions in Delaware and Maryland. In order to determine how well AWSC junctions would function if they were converted to roundabouts, a decision network was created as part of a knowledge-based expert system based on the research' conclusions. The results of the comparison analysis are consistent with those of several other studies, which found that given a comparable set of traffic circumstances and geometric restrictions, AWSC crossings consistently perform worse than signalised or roundabout types of intersections.
A study was done by Mahamoud Owais, et al. (3) to determine the service provided by two real roundabouts in the cities of Jeddah and Al-Madinah throughout time. The objective is to analyse and evaluate the level of performance of those two intersections within the case of already existing roundabouts and their parallel solution of signal controlled intersections in order to determine when it is appropriate to convert the current rotary intersection into a signal controlled intersection. Sidra and Synchro explain how traffic typically increases over time as a result of the simulation tools being given artificially fabricated demand scenarios. The underperforming year for the junctions described above is clearly identified by the results. Additionally, the impact of the rise in the number of left-turning vehicles is assessed for both categories. The suggested approach is believed to make it easier for practitioners to

continuously assess the benefit of switching from roundabout solutions to signal controlled intersections while also determining the span of service.
According to yield management, two- and four-way stop management, and signal management under a variety of traffic scenarios using the SIDRA software, Virginia P. Sisiopiku and Heung-Un Oh (4) compare the performance of roundabouts with four leg intersections. The study includes variations in lane widths, volume levels, turning volume splits, and variety of approach lanes. The analysis's findings suggest that roundabouts are the sole alternative for crossroads with multilane approaches that include a considerable amount of through and/or left turning traffic. Roundabout performance in comparison to signalised intersection performance at one-lane approaches and high traffic loads. For any percentage of left-turning traffic volume, roundabout capacities were shown to be higher than those of signal controlled junctions with two- and divided approaches. According to the study, roundabouts function better than controlled intersections, making them a desirable alternative.
Roundabouts have advantages over signalised control in some situations, according to Johnson, M. T. and Hange, W. A. (5). We compare the various operational properties of roundabouts and signalised control. Next, it is discussed how traffic engineers can recognise the usage of variations in roundabout and traffic light operation to comprehend safety and effective traffic operation, limit congestion, and identify the negative effects mostly associated with roadway and highway projects.
In order to make one- and two-lane roundabouts accessible to drivers with vision impairments, Bastian J. Schroeder et al. (6) investigated the use of pedestrian signalization options for crosswalks. Due of concerns about the security of roundabouts for blind walkers, auditory pedestrian signals have the potential to properly control how people and cars interact at these facilities. Due to the possibility of queue spillback into the circulating lane and delays to vehicle traffic, the use of pedestrian signals at roundabouts is controversial. The analysis, which uses a calibrated micro simulation model, evaluates creative approaches to crossing geometry and phasing at one-lane and two-lane roundabouts. The findings imply that roundabout pedestrian signals have a greater influence when vehicle levels approach their capacity, but that vehicle delay and queuing are frequently reduced by creative signal layouts. The findings are significant given current discussions about roundabout accessibility for walkers with vision impairments, which would eventually lead to a demand for signalization surely facility kinds.

4 Study of Location

4.1 Location Analysis

Because so many institutions and colleges are close by, Laxmi Mill Junction is regarded as one of the key intersections in Coimbatore City. Additionally, this intersection is situated closer to the bus stop and railroad station. Additionally, it links Ramanathapuram to Gandhipuram in a north-south direction and the Anna Salai to Nava India in an east-west direction. The key areas in Coimbatore city are these four areas.

Therefore, especially during peak hours, there is a considerable volume of traffic at this intersection. One of Coimbatore's busiest intersections is this one. To manage traffic flow, the junction has signals on each of its four sides. The traffic flow in this junction is occasionally managed by traffic police as well. The tranquil duration of the vehicle is particularly high at this intersection because of the large number of vehicles crossing.

4.2 Signal Analysis

Automatically operated Signals are provided in all the four directions. Sometimes the traffic control person controls the signals based on the traffic flow. Table 1 represent the signal timing during peak hours.

Table1. Peak hour Signal timing

Direction	Max Red time	Max Green time	Amber time
East	136	62	3
West	124	48	3
North	146	32	3
South	136	32	3

4.3 Data Collection from Location

The data related to road and vehicles is necessary to design the median and roundabout. For that process the data collection is done by surveyor method. The following data were collected for both peak and non-peak hours.

Width of the current road, size of the existing medians, traffic flow data, vehicle size, minimum turning radius, vehicle speed, potential conflicts, and pedestrian data.

4.4 Details of Existing Road And Medians

At North and south side the total width of the road is divided for two-way traffic, each side of the road has two lanes with an addition of two-wheeler lane.

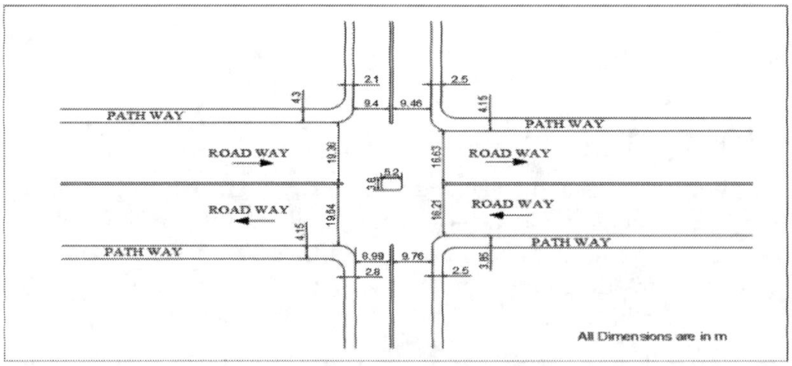

Figure 1 Width of Lanes

At East and west side the total width of the road is divided for two-way traffic, each side of the road has four lanes with an addition of two-wheeler lane. The Figure 1 shows the width of all the lanes and pathway.

Table 2 Width of existing medians.

Direction	Width of existing medians
East	0.6
West	0.6
North	0.85
South	0.6

5 Alternative for Signalized Intersection

5.1 Alternative A: Construction of Median with Pedestrian Bridge

Based on IRC recommendations the following three types of medians are proposed as an alternative for signalized intersection in that particular junction. Medians and the flow of vehicle from various directions is represented in figure 2 and 3.

- Central Median 1 &2
- Turning Median
- Distribution Median

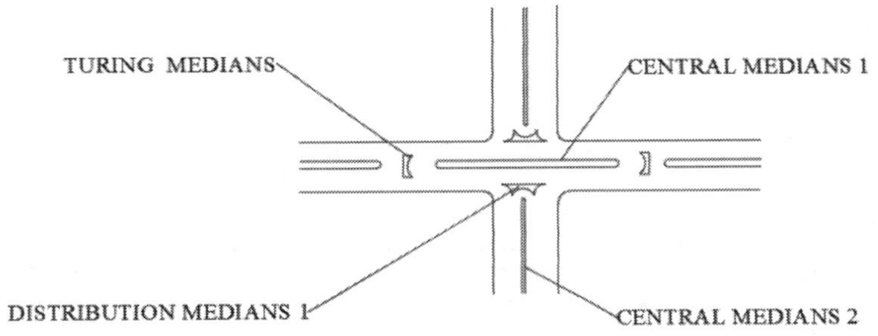

Figure 2 Medians Flow of Vehicle from Various directions

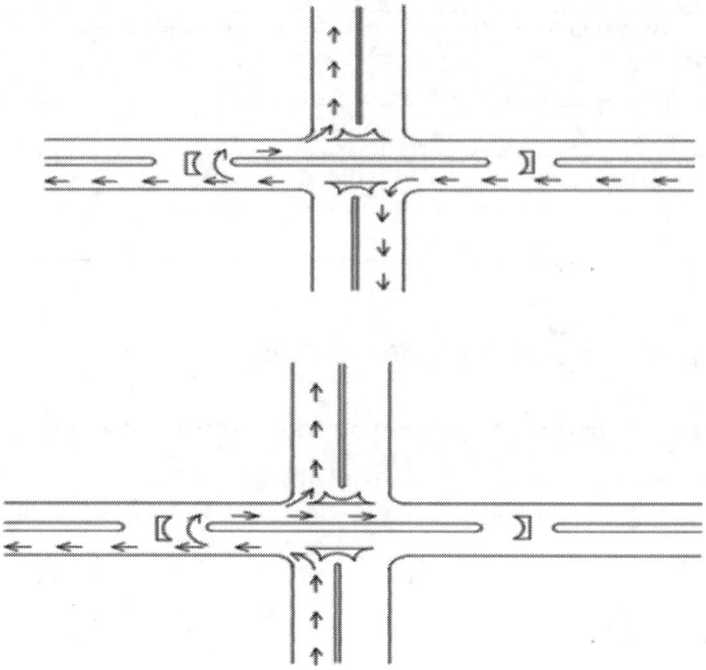

Figure 3 Flow of Vehicle from Various directions

5.1.1 Medians

In accordance with IRC Standards, the medians' width, length, and depth must be fixed and shown in table 3.

Table 3 Size of Medians

Type of Medians	Parameter	Dimensions
Central Median 1	Width	1.5m
	Length	210.36m
	Depth	1.0m
Central Median 2	Width	0.61 m
	Length	110.6m(Each side)
	Depth	1.0m
Turning Median	Width	7.0m
	Length	7.0m
	Depth	1.0m
	Radius of the turning curve	3.5m
	Distance from median	10.66m

Distribution Median	Top Width	10.56m
	Bottom Width	7.0m
	Length	7.0m
	Depth	1.0m
	Radius of the turning curve	3.5m
	Distance from median	10.66m

There are more pedestrians crossing the roads in all four directions at this intersection. In accordance with IRC requirements, a pedestrian crossing steel bridge has been designed in addition to the median to allow for pedestrian circulation.

5. 2 Alternative B: Construction of Roundabout with Pedestrian Bridge

An efficient Roundabout can handle 3000 vehicles per day from all directions. To collect the traffic flow data 12 surveys have been conducted, six surveys was conducted during peak hour traffic and the remaining six surveys was taken during non-peak hours traffic. Based on these 12 surveys, the conclusion was made that this intersection suitable for roundabout construction. If right turning traffic volume is more than 30%, we can strongly recommend roundabout for that particular location.
Building a roundabout requires a huge, wider area. Major district roads that are wide enough to support roundabouts at that specific site connect at the chosen intersection. The four directions of roads have unequal count of traffic, this unequal volume at four directions are also preferable to constructing roundabout. From the above mentioned things, we strongly agree and recommend constructing the rotary intersection at that particular intersection i.e. at laxmi junction.

5. 2.1 Design of Roundabout

The design of a roundabout encompasses the design of all of its different parts, including the radius of the central island, the waving width, the entering curve, and the exit curve. All the components are designed based on IRC recommendations. From design calculations and findings, the following values are adopted for roundabout at Laxmi mill junction.

Table 4 Design values of Roundabout

Parameter	Design values
Design speed	40 KMPH & Central
Shape	island with circular shape
Radius of Entry	e1 = 15 meters
Exit radius	1.5 x15 = 22.5 meters.
Radius of Central Island	1.33x15 = 19.95m
Weaving length & Angle	30 meters & 42 degree
Carriageway Width at entry and exit	e2 = 10 meters

Weaving section width	w = (e1+e2)/2 +3.5 =16 meters
Capacity of roundabout	North to south Qp = 3554.6 PCU/hr & West to South Qp = 3578.0 PCU/hr. West to North Qp = 3486.5 PCU/hr North to East Qp = 3502.5 PCU/hr

From the design let us consider the capacity of rotary intersection for the selected locality is 3486.5 PCU/hr. Similar to alternative A the Pedestrian crossing steel Bridge is designed in addition to the roundabout for pedestrian movement as per IRC recommendations.

6. Conclusions

We can infer from this analysis that creating a roundabout and a median is appropriate for the intersection in question. Although the designed roundabout may be able to manage current traffic as well as if there is a little rise in traffic flow rate, the designed median can handle the current traffic flow efficiently. There are many methods for managing the flow of traffic at at-grade intersections on roads. Being an unsignalized crossing, it contributes to the reduction of intersections in that area. When compared to signalised intersections, the travel time is smaller because the lane has enough space to move. The continuous flow of the vehicle is guaranteed from this median and roundabout design with the least amount of conflict sites. The safety of the vehicles as well as the pedestrian is improved than signalized intersections.

References

1. Frank gross, Craig Lyon "Safety effectiveness of converting signalized intersections to roundabouts", " *Accident Analysis & Prevention*" Volume 50, Pg.No 234-241, (2013)

2. Evdokia Vlahos "*Evaluating the Conversion of All-Way Stop-Controlled Intersections into Roundabouts*" "Journal of the transportation research board" Volume: 2078 issue: 1, page(s): 80-89, (2008)

3. Mahamoud Owais "*When to Decide to Convert a Roundabout to a Signalized Intersection: Simulation Approach for Case Studies in Jeddah and Al-Madinah*" "Arabian Journal for science and engineering" (2020).

4. Virginia P. Sisiopiku and Heung-Un Oh "*Evaluation of Roundabout Performance Using SIDRA*" "Journal of transportation Engineering" (2001).

5. Johnson, M T and Hange, W A "*Modern Roundabout Intersections: When To Use Them? A Comparison With Signalized Intersections*" "Institute of Transportation Engineers (ITE) 2003 Technical Conference and Exhibit" (2003)

6. Bastian J. Schroeder "*Toward Roundabout Accessibility—Exploring the Operational Impact of Pedestrian Signalization Options at Modern Roundabouts*" (2006).

7. "Indian Road Congress publishing the "*Manual of specifications and standards of at grade intersections*", pp107-178, (2010)

8. Douglas W. Harwood, Midwest Research Institute. Martin T. Pietrucha, Pennsylvania State University, and Kay Fitzpatrick, Texas University, "*Design of intersections on divided highways"* pp 34-76 (2012)

9. Ram Kumar.B, Adinarayana.B, Mane Sr Rohith, Aurora's Scientific Technological & amp; Research Academy released the paper "Analysis of Unsignalized intersection" pp 1-48, (2008)

10. Highway capacity manual, "*chapter 17- Unsignalized intersection*", pp 1-127 (2000)

11. State of Florida Department of Transportation. Released "*Median Handbook*" FDOT, pp 48-56, (2014)

12. Kay Fitzpatrick, Marcus A. Brewer, Willam L. Eisele Texas; Transportation Institute Developed the paper "*National Cooperative Highway Research Program*", pp 89-96, (2011)

13. Kadiyalai "*Traffic Engineering and Transportation planning*" Kanna Publishers.

14. Manisha Reji, Prem Reji Cherian, Reshma R, Sandeep Roy, International Research Journal of Engineering and Technology, released "*Design of Steel Fly-Over*", pp 1-34

15.P. Indhiradevi, P. Saravanakumar, S. Anandaraj, S.P. Narendran, P. Mohammedanas, S. R. Naveen, M. Nazir Ahamed, *"A case study of using median as an alternative for signal controlled intersection*", Materials Today: Proceedings, Volume 52, Part 3, 2022, Pages 1991-1996, (2021).

Evaluating the Environmental Sustainability of Electronic Products: A Case Study of Life Cycle Assessment In Coimbatore Region

Krithiga Palanisamy[1,*], *Sarathithasan*[2], *Sarulatha*[3], *Vaseeharan*[4]

[1]Assistant Professor, Department of Civil Engineering, Kongu Engineering College, Perundurai, Erode, Tamil Nadu, India

[2] Final Year Students, Department of Civil Engineering, Kongu Engineering College, Perundurai, Erode, Tamil Nadu, India

[3] Final Year Students, Department of Civil Engineering, Kongu Engineering College, Perundurai, Erode, Tamil Nadu, India

[4] Final Year Students, Department of Civil Engineering, Kongu Engineering College, Perundurai, Erode, Tamil Nadu, India

Abstract: The growth of science and technology is to lead an increasing quantity of electrical and electronic equipment trash (E-waste); electronic products became important and necessary products in our daily life. Due to its quick volume increase and toxic nature, electronic waste has turned into an environmental problem. E-waste treatment is a more troublesome task than handling electronic waste products that lacks a clear direction, which results in land filling, disposal, etc. E-waste contains several dangerous components that could harm the atmosphere and human health. This research attempt to the dismantling and separation of the e-waste components (mobile phone, laptop, desktop). In addition, calculate the life cycle GHG of e-waste. To evaluate the impact categories with the help of openLCA software. Then the Ecoinvent v3.6, cut off and ReCiPe 2016 Midpoint (H) indicators were used. The greenhouse is emission is mainly based on the transportation. Finally, this paper shows the financial value and the environmental effect of present disposal of e-waste. Under e-waste running most domestic e-waste kept at home because people don't know where to discard. The present practice adds more greenhouse gas to the atmosphere because of the distance between the dismantling unit and the recycling unit.

Keywords: E-waste, Life Cycle Assessment, Life Cycle GHG, Impact categories, open LCA.

1. Introduction

Electronic wastes are the residues from rejected electrical or electronic products like computers, mobile phones, TVs, motors, freezers, washing machines, and air conditioning units. Approximately 17.4% of the e-waste that was formed worldwide in 2019 was disposed of or recycled effectively. The destiny of the lasting 82.6 percent, that might be disposed of without adequate handling or recycling, was not known. Due to the high demand for electronic goods in modern society, e-waste production has emerged as one of the major problems. The explosive growth of computing and networking devices among other electrical consumables is mostly to blame for the rise in e-waste output globally (Mmereki et al .,2016). Metals make up 60% of e-waste, tracked by plastics at 15% and metal-plastic mixtures at 5%. When the attention of poisonous metals including lead (Pb), mercury (Hg), arsenic (As), cadmium (Cd), selenium (Se), and chromium (Cr) in the environment exceeds the allowable limit, it may cause risk and dangers to people and the ecology through straight or indirect contact. The three primary sources of e-waste are consumer electronics (13.7%), info and transportations technical apparatus (33.9%), and

*Corresponding author: krithipalanisamy@gmail.com

© The Authors, published by EDP Sciences. This is an open access article distributed under the terms of the Creative Commons Attribution License 4.0 (https://creativecommons.org/licenses/by/4.0/).

waste household equipment (approximately 45%). According. to the United States Environmental Protection Agency (USEPA), eight metals—Pb., Cr, a As, Zn., Cd, Cu., Hg, and Ni—are the most common and ecologically toxic (Chakraborty et al.,2022). Given that the majority of electronic trash includes toxic materials and metals, managing this garbage is one of the major difficulties facing the globe today. E-waste makes up a relatively modest percentage of all municipal solid garbage. However, the population and per capita electronic item consumption impact the quantity of e-waste generated (Shittu et al.,2021). The production of electronic waste has significantly expanded as a result of rapid automation and the accessibility of high technology. The International Solid Waste Association (ISWA) estimates that 53.6ametric tons or so of electronic waste were produced globally in 2019.

1.1 GLOBAL E-WASTE

Due to the population's rapid growth, electronic waste from electrical and electronic equipment is produced globally. Electronic wastes are the residues from rejected electrical or electronic products like computers, mobile phones, TVs, motors, freezers, washing machines, and air conditioning units (Andeobu et al.,2021). Every year, more and more electronic waste is produced, and recycling is not keeping up with this growth. E-waste is frequently disposed of by landfilling, which is a widespread activity that is both clear and familiar (Andooz et al.,2021). Globally, 53.6 million tonnes (Mt) of e-waste were produced in 2019. Only 17.4% of the total volume of e-waste was recycled or appropriately disposed of. It was unknown what would happen to the remaining 82.6 percent, which might be disposed of without proper processing or recycling (zuo et al.,2020). Three times more e-waste was produced globally in 2019 than there were people on earth. It is predicted that more than 74 Mt would have been produced in e-waste globally by 2030. The two main problems are that manufacturers and sellers of electrical and electronic equipment (EEE) are not held accountable for the disposal of their products after their useful lives (EoL), and there is a lack of efficient infrastructure for the collection and recycling of e-waste in both developed and developing countries (Bhaskar and Kumar., 2019). In 2006, the UN estimated that 50 million metric tonnes of electronic waste were discarded annually worldwide. Other nations, like India, the amount of e-waste created over the next ten years might increase by as much as 500 percent, conferring to a UNEP statement titled "Recycling – from electronic waste to Resources." This includes computers and mobile phones. With nearly 3 million tons yearly production of electronic waste, the United States leads the globe in this category. 2.29 million tons produced by China at home, second only to the United States (estimate from 2010). China also continues to be a significant e-waste removal ground for industrialized nations despite having outlawed the import of e-waste. Today's society is centred around technology, and by always seeking out the latest and most advanced items, we are creating a significant amount of E-waste (Rajesh et al.,2022). According to estimates, India and Thailand produced 0.33 and 0.1 million tonnes of e-waste, respectively, in 2007 (Ahirwar and Tripathi., 2021). However, when comparing consumption per capita, Europe produces 16.2 kg, whereas America generates 13.3 kg, Asia produces 5.6 kg of meat per capita (Ahirwar and Tripathi., 2021). Cell phones have surpassed all other e-waste product categories since the introduction of the iPhone. Hazardous, valued, and rare materials can be found in electrical waste. Multipart electronics can contain up to 60 components. Apple sold more than 796 million I Devices as of 2013. (iPod, iPhone, iPad). To encourage customers to buy new phones, cell phone manufacturers produce short-lived phone. Businesses give these products such limited shelf lives since they are aware that consumers will need and purchase an invention if it is produced (sharma and jain.,2020). Figure 1 shows the global forecasting of e-waste in million metric tons.

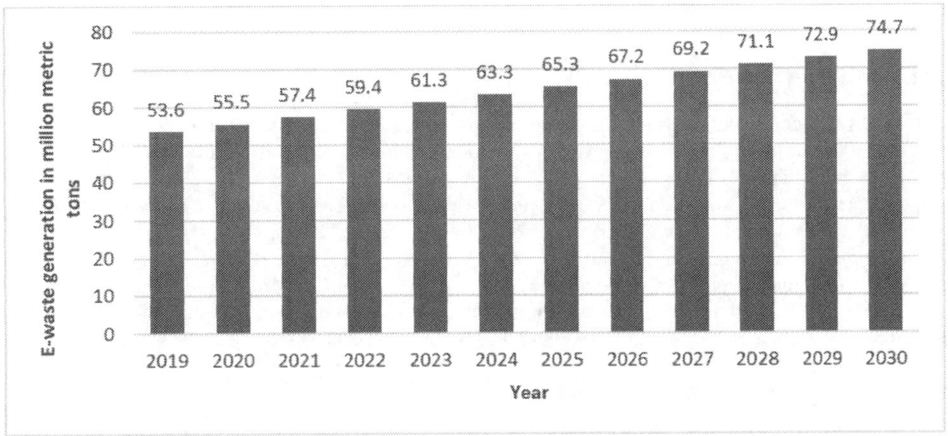

Figure 1 Global Forecasting of E-Waste (Forti et al.,2020)

1.2 E-WASTE IN INDIA

India now generates more than 3.23 MT of e-waste yearly, ranking third in the world behind China and the United States, according to the Global E-Waste Monitor Report 2020 (Forti et al., 2020). India's manufacture of e-waste increased by over 43% between 2018 and 2020 (Sanito et al.,2021). By 2050, it is anticipated that the nation will produce over 161 million tonnes of electronic trash (Yang et al.,2021). In addition to the considerable volumes of e-waste produced within the nation, it is also known that developed countries import significant amounts of e-waste (Sengupta et al.,2022). India produces a significant amount of e-waste, however compared to other countries which also produce e-waste, the country's collection and recycling rates are still quite low. This shows that electronic waste management in this developing nation is still in its beginning (Intrakamhaeng et al.,2020). However, the most recent findings show that only 5 to 10% of the entire amount of e-waste formed in India was sustainably recycled by the official sector using trying to cut equipment, with the remainder being treated by the unregistered familiar sector (Ilankoon et al.,2018). In India the quantity of e-waste produced varies by government. The three states with the highest levels of electronic waste are Maharashtra, Tamil Nadu, and Andhra Pradesh. Additional conditions that generate important amounts of e-waste include Uttar Pradesh, West Bengal, Delhi, Karnataka, Gujarat, Madhya Pradesh, and Punjab. E-waste is also excessively produced in built-up areas; 65 Indian cities account for extra 60% of total e-waste production in India. After Mumbai, the next-largest producers of e-waste are Delhi, Bangalore, Chennai, and Kolkata (Xavier et al.,2021). Figure 2 shows the percentage of e-waste generated in India.

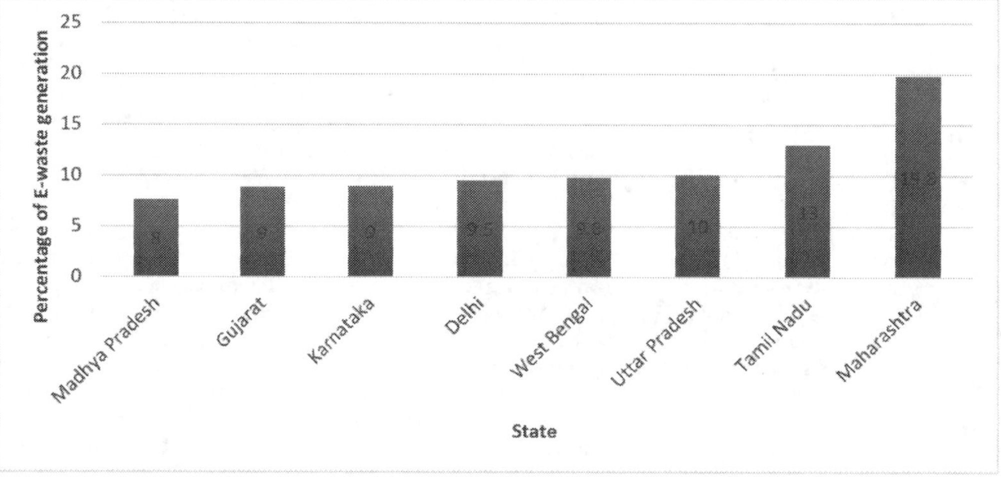

Figure 2 Percentage of E-waste Generated in India (Forti et al.,2020)

1.3 LCA OF E-WASTE

The LCA has been employed in various studies to generate ideas for management improvement from an environmental standpoint and to evaluate the influences related to the end of life of WEEE. Some studies reported the advantages of material treatment and recovery as well as their effects on various environmental aspects. Other studies demonstrate that reusing typically has a smaller impact when compared to other options like landfill or incineration (Pérez-Martínez et al., 2021). The collection and logistics of WEEE were also studied because of their significant variability and ability to impair their environmental benefits. Some of these studies examine landfill, recycling, and incinerator practises for WEEE from computers. The bulk of these research show that although recycling has negative impacts on the environment, such effects are balanced by environmental advantages of the improved materials (Bressanelli et al.,2020). Some experts believe that raising WEEE recycling rates or changing the materials used to make equipment could have a significant positive effect on the environment (Sanito et al., 2021).

Generally, these educations results suggest that the LCA is a beneficial tool for conducting a comprehensive environmental assessment of final disposal scenarios. Reusing WEEE is better for the environment than other forms of treatments that are now more fashionable, such recycling, burning, or landfilling. Reuse often outperforms other options from an environmental perspective, as according to studies

(Patil et al., 2020). Other works concentrate on specific electrical and electronic devices (EEE). When three end-of-life scenarios for air compressors have been evaluated, taking into consideration their whole life cycle, it was discovered that remanufacturing was the best course of action for all impact categories considered (Milutinovic et al.,2017). The lifespan is cited as a crucial element in the effectiveness of reuse in many of these studies that focused on Information and Communication Technologies (ICT) equipment. But a small number of studies have so far conducted an extensive environmental assessment of WEEE managing through repurposing (Mangmeechai., 2022). Finally, some focused on mobile phones and came to the conclusion that for used smartphones, repurposing is a better management option that refurbishing.

2. Materials and Methods

2.1 Materials Used

Telecom electronic devices were selected for dismantling, the selected electronic waste components were desktop,laptop and mobile phones.

Table 1 E-waste components

S.no	Components	Weight of components in kg
1.	Desktop	2.76
2.	Laptop	1.96
3.	Mobile phones	0.81

Figure 3 E-waste components-desktop, laptop, mobile phones

The dismantling process were done in the dismantling unit. To disassemble the waste, they employ simple tools like hammers, spanners, pliers, screwdrivers, and drills. By using the tools, they dismantle the e-waste components to separate the metals like Aluminium, Copper, Steel, Plastic, Glass, Pcb, Battery and Others. Then, each part is evaluated to. analyse the proportion of waste in the main components.

Figure 4 Metals present in E-waste components

2.2Methods

2.2.1openLCA

openLCA is a specialized Life Cycle Assessment (LCA) and Footprint software program that is free to use. It was developed by Green Delta and has a wide range of features. A piece of open source software is openLCA. LCA, or life cycle assessment, was first the principal use of openLCA. Later, the capabilities of openLCA were expanded to include models for economic life cycle assessments, particularly when combined with LCA in the form of life cycle costing. The desktop programme openLCA functions without an online connection. Additionally, a web-based database that may be used for data and model exchange can be setup to work with openLCA.

The method used in openLCA software was ReCiPe 2016 midpoint(H) and distribution default and system processes were applied to study the life cycle GHG. Primary data, e.g., e-waste components, Transportation And secondary data e.g., openLCA software (agribalyse_v301 database) used.

2.3Study Area

Green India Recyclers in Coimbatore, is an electronic waste management company. The guiding principle of Green India Recyclers is pretty straightforward. In India, big cities like Coimbatore are developing into e-waste centres. This is a severe issue that needs to be resolved right away. particularly in urban areas like Coimbatore, which has a large number of heavy manufacturing sectors, IT firms, as well as several hospitals, schools, colleges, and businesses. In such big cities, the amount of electronic equipment that is utilised and thrown out increases daily. They are dedicated to eliminating, reducing, and recycling electronic waste. In addition to helping businesses and individuals with safe e-waste disposal, they will also teach them how to reduce their own e-waste footprint. Furthermore, by recycling products, we are able to further reduce e-waste generation. Figure 3 shows the collected waste placed in dismantling unit.

Figure 5 Collected waste placed in dismantling unit

3.Result and Discussion

3.1 Flow of E-waste form household and industry

Industrial e-waste is not a concern because most firms have extremely defined e-waste management practises in place. These industries convey their e-waste to professional disassembly sectors. Households, on the other hand, lack any instructions or regulations on e-waste. The recycling and gathering system is not effectively run. Owners can dispose of their outdated e-waste items by selling them to distributors, who then Trade them to e-waste dealers. E-waste can also be disposed of with regular household solid waste. After segregation of solid waste, the e-waste is separated and collected by the dismantler. Now the Industrial e-waste and the household. e-waste were collected at the same point dismantling unit where the usable products were resale and refurbish and sold to the local electronic shop. Finally, the extracted metals sent to the recyclers for recycling.

3.2 E-waste components

Three types of electronic waste components were selected and pull to pieces. The separation method done in formal dismantling centre. The e-waste components are short in Table2.

Table 2 E-waste components (g per unit)

Components	Glass	Plastic	Steel	Copper	Aluminium	PCB	Battery	Others	Total
Desktop	725	560	1135	50	30	225	-	35	2760
Laptop	290	995	115	40	125	280	-	75	1960
Mobile	250	220	75	-	30	110	120	5	810

Usually, landfill sites do not include waste in its original form, according to local authorities. Plastic, aluminium, steel, and copper may all be recycled locally. Landfills get the glass from desktop, laptop, and mobile phone shades that comprise lead and plastic (such as the front and back plastic cases of mobile phones). After the dismantling process the metals obtained were sent to the recyclers where the metals were recycled and the circuit board undergo a process of cleaning and removing the lead present in it then the circuit board is grinded to the powder format used as the fibre material . About 75% of the weight of desktop, laptop, and mobile phone components were recyclable parts, correspondingly. Table 3 lists non-recyclable electronic waste. Laptops produced the least amount of garbage, while mobile phones produced the most.

Table 3 Discarded waste from components (g per unit)

	Desktop	Laptop	Mobile
Waste	760	365	255
% By weight	27.5%	18.6%	31.4%

3.3.Life Cycle Green House Gas of E-waste

The life cycle GHG of each type of e-waste depends upon transport distance. Here the components are dismantled in the same unit the transportation distance from the house and the dismantling unit is 80-90 Km and the distance between the dismantling unit and the recycling unit is approximately 520-530 Km. E-waste administration involves collecting , take apart, transporting , recycling, and disposing of residue.

First the telecom device desktop, laptop, and mobile phones are transported to the dismantler there the metal extraction process is done. The metals are transported to the recycling plant. Plastic, aluminium, copper, steel are the valuable metals present the waste components that undergo the recycling process the valuable metals are used as the raw materials. The Printed Circuit Board undergo the special recycling process the other metals presented in the board are removed and the empty board is grinded using the machines the final output of board will be in powder format it will be used as fibre material. The Life Cycle GHG is complex to transportation. Here we used two software's to calculate the GHG emission from the components.

3.4 Open LCA

Figure 6 openLCA software

openLCA is a professional Life Cycle Assessment (LCA) and Footprint software that is free to use. It was developed by GreenDelta and has a wide range of features. A piece of open source software is openLCA. LCA, or life cycle assessment, was first the principal use of openLCA. Later, the capabilities of openLCA were expanded to include models for economic life cycle assessments, particularly when combined with LCA in the form of life cycle costing.
Here for using openLCA software the first step is to feed the input details as the weight of the component and transportation distance. The needed output is the impact created by the obtained metals present in the component. Next the impact assessment method selected for the software is ReCiPe 2016 midpoint(H). From this method the software attains many impact categories in that we selected global warming. Finally, the impact categories were listed from that the global warming is considered. Table 4 shows the results obtained from both of the software's

Table 4 Life Cycle GHG of E-waste(kgCO_2eq/unit)

E-waste components Software used	**Desktop**	**Laptop**	**Mobile phones**
openLCA	3.30974E+01	3.31E+01	1.65590E+01

From the results the Mobile phones is considered to the less emission the greenhouse has and the laptop considered to the high emission of greenhouse gas. The emission factor is mainly based on the transportation process and the weight of the metals present in the components.

Cell phones, desktop computers, laptops, and other sorts of e-waste were the focus of the education. According to the worth of its parts, the e-waste estimations shown overhead can be classified into two groups: (1) Unrecyclable or waste materials. For example, items that have been dumped in landfills, such as plastic and glass with lead gratified (TV screens), are not recyclable. which made up 30% of the overall amount of e-waste produced; (2) recyclable components that may be used as raw materials, such as plastic, steel, copper, and aluminium, which made up 70% of the total.

Similar findings were made by some studies, namely that reusing is not as environmentally helpful as one might anticipate, particularly in light of the impact of fossil fuels utilised in transportation. The transportation lengths from waste generation to disposal locations and recycling businesses are the key variables that determine life cycle GHG. The emissions from transportation make up a sizeable component of the total life-cycle emissions when there is improper organisation between the stages.

A more effective e-waste gathering system is required, and collection standards and locations will impact revenue and GHG emissions. As a result, regional collecting, sorting, and dismantling centres must be built between recycling facilities and dismantling facilities. Additionally, the travel distance between each facility shouldn't be higher than 200 kilometres. Given that e-waste contains a significant number of precious metals compared to mineral ores, it should be seen as a potential opportunity to recycle or recover valuable metals (such as copper, aluminium, and steel). Organizations might benefit from taking into account public knowledge of the human risk involved in e-waste recycling activities in terms of lowering potential negative effects on society. To convince consumers to dispose of their electronic waste at collection sites rather than mixing it in with their regular household garbage, incentives may be necessary.

4. Conclusion

Both useful elements and harmful compounds can be found in abundance in electronic garbage. One of the most pressing issues affecting the world now is the management of electronic garbage. The effectiveness of electronic waste management practises in terms of the environment can be measured using life cycle assessment, an efficient and widely used environmental management tool. There must be more dismantling and recycling unit so that the transportation distance will reduce and the informal dismantling unit will be reduced. In order to obtain more complete results, 40% of the studies taken into consideration also finalised their findings using additional environmental techniques such as material flow analysis, multi-criteria decision analysis, energy analysis, and risk assessment.

References

1. Ahirwar, R. and Tripathi, A.K., 2021. E-waste management: A review of recycling process,environmental and occupational health hazards, and potential solutions. Environmental Nanotechnology, Monitoring & Management, 15, p.100409.
2. Andeobu, L., Wibowo, S. and Grandhi, S., 2021. An assessment of e-waste generation and environmental management of selected countries in Africa, Europe and North America: A systematic review. Science of The Total Environment, 792, p.148078.
3. Andooz, A., Eqbalpour, M., Kowsari, E., Ramakrishna, S. and Cheshmeh, Z.A., 2021. A comprehensive review on pyrolysis of E-waste and its sustainability. Journal of Cleaner Production, p.130191.
4. Bhaskar, K. and Kumar, B., 2019. Electronic waste management and sustainable development goals: Is there a business case for linking the two?. Journal of Indian Business Research.
5. Bressanelli, G., Saccani, N., Pigosso, D.C. and Perona, M., 2020. Circular Economy in the WEEE industry: A systematic literature review and a research agenda. Sustainable Production and Consumption, 23, pp.174-188.
6. Chakraborty, S.C., Qamruzzaman, M., Zaman, M.W.U., Alam, M.M., Hossain, M.D., Pramanik, B.K., Nguyen, L.N., Nghiem, L.D., Ahmed, M.F., Zhou, J.L. and Mondal, M.I.H., 2022. Metals in e-waste: Occurrence, fate, impacts and remediation technologies. Process Safety and Environmental Protection, 162, pp.230-252.
7. Forti, V., Balde, C.P., Kuehr, R. and Bel, G., 2020. The Global E-waste Monitor 2020:Quantities, flows and the circular economy potential.

8. Ilankoon, I.M.S.K., Ghorbani, Y., Chong, M.N., Herath, G., Moyo, T. and Petersen, J., 2018. E-waste in the international context–A review of trade flows, regulations, hazards, waste management strategies and technologies for value recovery. Waste Management, 82, pp.258-275.
9. Intrakamhaeng, V., Clavier, K.A., Liu, Y. and Townsend, T.G., 2020. Antimony mobility from E-waste plastic in simulated municipal solid waste landfills. Chemosphere, 241,p.125042.
10. Mangmeechai, A., 2022. The life-cycle assessment of greenhouse gas emissions and life-cycle costs of e-waste management in Thailand. Sustainable Environment Research, 32(1),pp.1-13.
11. Milutinović, B., Stefanović, G., Đekić, P.S., Mijailović, I. and Tomić, M., 2017.Environmental assessment of waste management scenarios with energy recovery using life cycle assessment and multi-criteria analysis. Energy, 137, pp.917-926
12. Mmereki, D., Baldwin, A. and Li, B., 2016. A comparative analysis of solid waste management in developed, developing and lesser developed countries. Environmental technology reviews, 5(1), pp.120-141.
13. Patil, R.A. and Ramakrishna, S., 2020. A comprehensive analysis of e-waste legislation worldwide. Environmental Science and Pollution Research, 27(13), pp.14412-14431.
14. Pérez-Martínez, M.M., Carrillo, C., Rodeiro-Iglesias, J. and Soto, B., 2021. Life cycle assessment of repurposed waste electric and electronic equipment in comparison with original equipment. Sustainable Production and Consumption, 27, pp.1637-1649.
15. Rajesh, R., Kanakadhurga, D. and Prabaharan, N., 2022. Electronic Waste: A critical assessment on the unimaginable growing pollutant, legislations and environmental impacts. Environmental Challenges, p.100507.
16. Sanito, R.C., You, S.J. and Wang, Y.F., 2021. Application of plasma technology for treating e-waste: A review. Journal of environmental management, 288, p.112380.
17. Sanito, R.C., You, S.J. and Wang, Y.F., 2021. Application of plasma technology for treating e-waste: A review. Journal of environmental management, 288, p.112380.
18. Sengupta, D., Ilankoon, I.M.S.K., Kang, K.D. and Chong, M.N., 2022. Circular economy and household e-waste management in India: Integration of formal and informal sectors. Minerals Engineering, 184, p.107661.
19. Sharma, K.D. and Jain, S., 2020. Municipal solid waste generation, composition, and management: the global scenario. Social Responsibility Journal, 16(6), pp.917-948.
20. Shittu, O.S., Williams, I.D. and Shaw, P.J., 2021. Global E-waste management: Can WEEE make a difference? A review of e-waste trends, legislation, contemporary issues and future challenges. Waste Management, 120, pp.549-563.
21. Xavier, L.H., Ottoni, M. and Lepawsky, J., 2021. Circular economy and e-waste management in the Americas: Brazilian and Canadian frameworks. Journal of CleanerProduction, 297, p.126570.
22. Yang, Y., Liew, R.K., Tamothran, A.M., Foong, S.Y., Yek, P.N.Y., Chia, P.W., Van Tran, T.,Peng, W. and Lam, S.S., 2021. Gasification of refuse-derived fuel from municipal solid waste for energy production: a review. Environmental chemistry letters, 19(3), pp.2127-2140.
23. Zuo, L., Wang, C. and Sun, Q., 2020. Sustaining WEEE collection business in China: The case of online to offline (O2O) development strategies. Waste Management, 101, pp.222-230.

Evaluating the Properties of Self-healing Concrete: A Review

Pooja Damodaran[1], *Lakshmi* Thangasamy[2*], and *Jaganathan* Venugopal[3]

[1]Research Scholar, Department of Civil Engineering, Saveetha School of Engineering, Saveetha Institute of Medical and Technical Sciences, Chennai, India - 602 105.
[2*]Associate Professor, Department of Civil Engineering, Saveetha School of Engineering, Saveetha Institute of Medical and Technical Sciences, Chennai, India - 602 105.
[3]Assistant Professor, Department of Civil Engineering, Saveetha School of Engineering, Saveetha Institute of Medical and Technical Sciences, Chennai, India - 602 105.

Abstract. The development of cracks is a regular phenomenon that causes water and various chemicals to seep into the structure reducing the strength and longevity of the building. The intrusion of external materials might affect the reinforcements when exposed to moisture and CO_2 exposure, degrading the structure. Regular maintenance and treatments are needed to fix the cracks and maintain the structure. Research is in progress identifying solutions to these issues with active and self-processes in concrete which aids in repairing the cracks by generating calcium carbonate ($CaCO_3$) particles which could seal the porosity of concrete and micro-cracks. Various bacteria like Bacillus pasteurii, Bacillus subtilis, Bacillus cohnii and Bacillus sphaericus are chosen for the crack sealing treatment as those bacteria could survive in an alkaline medium. Distinct species of bacteria have various growth conditions. Bacterial self-healing treatments were situated in an environment containing various minerals and chemicals at a constant temperature and for a specific amount of time to develop healing properties. This paper investigates the different experimental works studied on self-healing processes in large concrete samples with different sizes and materials. Studies revealed that bacteria enhanced the systemic toughness, flow ability, durability, and flexural modulus of normal concrete. This paper also investigates the different mathematical formulations established to examine the stress-strain behaviour of bacterium, which has been utilized to increase the strength of concrete.

Keywords- *Bacillus pasteurii, Bacteria, Concrete, Stress- strain Bacillus sphaericus*

1 Introduction

Concrete is a necessary building material widely used to construct roads, buildings, tunnels, subways, bridges, and other structures. Based on the cracks and precautions in concrete frequently form owing to tensile forces. These cracks often spread further as they improve the results of different environmental conditions, which raise the expense of repairs.

* Corresponding author: lakshmits.sse@saveetha.com

© The Authors, published by EDP Sciences. This is an open access article distributed under the terms of the Creative Commons Attribution License 4.0 (https://creativecommons.org/licenses/by/4.0/).

Crystalline admixture repairs cracks once bacteria and calcium are added, improving compressive strength and toughness. It will be required to patch any cracks that have formed because of the concrete's increased permeability, which allows water to pass across it and come into contact with the encouragement of the structure. This causes degradation to begin over time, reducing the toughness of the building. Calcareous crystals form when organisms are introduced into concrete, obstructing the microscopic pores and fractures [1, 2]. The capacity of microorganisms to live in a highly alkaline determines which germs are chosen. Most creatures perish in a microenvironment with a pH value of 10 or above [3]. When the pH of the highly alkaline material falls between 10 and 11.5, the bacterial spores are released. As indicated in several bacteria (B. pasteurii, Deleya Halophila, Halomonasrurihalina, Myxococcus Xanthus, B. megaterium, Bacillus subtitles, B. sphaericus, Thiobacillus) besides Bacillus that can survive in an alkaline condition [4]. Based on the size of the materials and cracks due to the primary importance of selected self-healing techniques and methodology of materials, however, encapsulation of the chemical composition of healing agents shows better results due to effective reaction between un-hydrated particles and healing materials. For the investigation of bacteria, several species of bacteria were utilized by various scientists. For example, [5] employed Bacillus cohnii microbes to precipitate $CaCO_3$, [6], [7], and [8] employed Bacillus pasteurii, while Dick et al. [9] used Bacillus lintus.

$$CO(NH_2)_2 + H_2O \longrightarrow NH_2COOH + NH_3 \quad (1)$$
$$NH_2COOH + H_2O \longrightarrow NH_3 + H_2CO_3 \quad (2)$$
$$H_2CO_3 \longleftrightarrow HCO^-_3 + H^+ \quad (3)$$
$$2NH_3 + 2H_2O \longleftrightarrow 2NH_4^+ + 2OH^- \quad (4)$$
$$HCO^-_3 + H^+ + 2NH_4^+ + 2OH^- \longleftrightarrow CO_3^{2-} + 2NH_4^+ + 2H_2O \quad (5)$$

Although microorganisms have polarized cell walls, they take metal ions to form their surroundings, such as Ca_{2+}, to deposit on their cell surfaces. The ensuing reaction between the Ca_2^{+}-ions and the CO_3^{2-}ions results in the precipitate of $CaCO_3$ at the cell membrane, which acts as a crystalline phase (Equation (6)). [10]

$$Ca_2^+ + CO_3^{2-} = CaCO_3\downarrow \quad (6)$$

The result of this study was that Bacillus cereus chose to give the most compressive strength, and S.pasteurii gave the highest decreases in digestion. Bacteria might survive in extremely alkaline conditions by forming spores similar to plant seeds. The spores have a very strong wall and become active when the concrete begins to crack and water enters the building. In this research, the previous researchers to develop the mechanical strengths and percentage of increasing strength were used, and the different types of Encapsulation techniques with furthermore of microbial species in Table 1.

Table 1 Analysis of concrete mix, bacteria species, and strength of samples

References	Grade of concrete	Encapsulation technique	Bacteria species used	% increase in strength
[11]	M30	Aerated concrete granules	Bacillus cohnii,	25
[12]	M35	Air-entraining admixture	Bacillus cohnii,	33

[13]	M35	Ceramsite	Bacillus subtilis	26
[14]	M40	Expanded perlite	Bacillus cohnii	27
[15]	M35	Natural fibers	Bacillus subtilis, Bacillus cohnii, Bacillus sphaericus,	42
[16]	M30	Polyurethane	Bacillus subtilis	29

2 Materials and Methods

2.1 Materials

Crack healing in bacterial concrete is thought to occur via the metabolic processes of transforming salt to carbonate, which results in crack sealing. This biologically mediated process effectively sealed sub-millimeter (0.05 mm to 0.87 mm [17], 5 to 10 mm [18 – 19], 100 mm [20], 200 mm [21], 205 mm [22], and 300 mm [23] cracks. Further development of this new type of concrete that heals itself is expected to result in a more durable and environmentally friendly concrete that can be especially suited to uses in humid conditions where corrosion in reinforcement frequently hinders the durability of conventional constructions made of concrete.

2.2 Bacteria concrete

Previous research used Bacillus pasteurii, Bacillus aerius, Bacillus sphaericus, Sporosarcina aquamarine, Bacillus megaterium, and other urease bacteria are commonly proposed for self-healing concrete. Bacillus pasteurii, a Gram-positive bacteria soil bacterium, can grow normally at temperatures ranging from 15 to 40 degrees Celsius. Bacillus pasteurii has exceptional uric acid activity, which can rapidly decay urea in the surroundings into ammonia and carbon dioxide. Gram-positive bacteria, including Bacillus megaterium Bacterial urease, can hydrolyze urea, which will cause $CaCO_3$ precipitation and provide self-healing capacity in concretes [24]. The investigation employed B. sphaericus LMG 225 57 (BCCM, Gent) [25, 26]. Previous research indicates that it is frequently used in the deep sea due to its capacity to produce highly resilient spores in extreme conditions. Its temperature range for survival and growth is mainly between 3 and 45°C [16]. B. sphaericus is a Gram-positive aerobic bacterium that produces elliptical spores and can convert urea-based to urease. B. sphaericus comes into touch with people when they improve in the new state and hardened state properties [27, 28]. According to studies conducted to induce the formation of calcium carbonate precipitates using Bacillus cohnii, Bacillus pasteurii, and Bacillus sphaericus, concerning the repaired width of bacteria, the Bacillus cohnii self-healing performance on concrete samples is improved.

Non-bacteria that induce calcium carbonate precipitate include Bacillus pseudofirmus, Bacillus cohnii, Bacillus halodurans, Bacillus mucilaginous, Enterococcus faecalis, Geobacillus stearothermophilus, Bacillus subtilis, and others. Bacillus subtilis is a Gram-positive bacterium that produces oval or cylindrical spores [29]. Because Bacillus subtilis is used in agriculture and some medicines, it is not harmful to humans. In an alkaline environment, the pH value of the solution increases with the addition of NH_3 and CO_2. The pH also affects the development of microorganisms. Each type of microorganism has a specific pH range. In a test tube, nutrients with pH values ranging from 4 to 12 were created.

The sample's turbidity was measured using a photo measurement device to conduct the test. It was discovered that the increase occurred best in the pH range of 7.5 to 9.0. Bacillus pasteurii and Bacillus sphaericus grew best in pH levels of 7- 9 [29].

2.3 Mesurement of samples

Supplementary cementitious materials provide long-term and performance benefits to those who construct and occupy buildings of all types. The increased use of these sustainable materials is partly due to enhanced workability and combining, flexural and compressive advantages, pump ability, durability against chlorides and sulfates, cooler temperatures for mass-concrete and alkali-silica reaction reduction, and reduced permeability. Cementitious mixes not only produce stronger, more durable, high-performance concretes, but they also help to reduce global climate impact by reducing the use of energy and emissions of greenhouse gases [30]. Addition of Bacteria Subtilis and cementitious materials cultivated and introduced to the water during the concrete's mixed process in different concentrations. Using regular Portland cement, fine aggregates, aggregates, and water to create specimens to research and evaluate the concrete's healing ability. The following moulds have been used: 160 mm × 160 mm × 70 mm, 150 mm × 150 mm × 600 mm, and 150 mm × 150 mm × 150 mm. The samples were stored in the space for 28 days between 20°C - 25°C. The produced 150 mm cube was put through a compressive load after 28 days, and it was discovered that the mean mechanical properties were 55.2 N/mm^2 with incorporates several of 2.19 N/mm^2.

2.4 Combined Ureolytic Bacteria

Bacterial urease can break down urea, precipitating calcium carbonate and self-healing capacity in concrete construction. This civilization was produced in a continuous reactor using activated biomass. It must have been crowded with 1 litre of activated sludge obtained from a waste disposal center oxygenation, followed by sediments in Oxidative damage by reactive oxygen emerges. It also contained 2 g/lt of the urease microbes that benefit ecologically from the reactor core's continual rotation and mixing at 100 rpm and 28°C, which promotes their development [31]. Urease creatures secrete urea hydrolysis-catalyzing urea in metabolic processes.

2.5 Density of Bacteria concrete

Lightweight concrete can be made from a lightweight aggregate, and the type produced is determined by the accessibility of the substance, as seen in expanded sand, shale rock, and smelted fly ash. LWA is also utilized to enhance the concrete's capacity for self-healing. Furthermore, organic substances such as pumice stone, volcanic scoria, volcanic slag, perlite, and tuff were utilized. In the meantime, lightweight concrete can be used for thermal insulation, masonry walls, and structural purposes. The concrete specimen's hardness will be impacted by replacing the standard 2 mm - 4 mm material with lightweight materials, identically sized material corresponding to a 15 kg/m^3 healing ingredient content. Concrete containing a liquid solution enclosed in LWA had a significantly better capability to repair fractures [31, 32].

3 Testing of self-healing concrete

The mechanical results of bacteria-modified concrete on different concrete specimens are discussed. After curing for 7 days, 28 days, and 90 days, tests were performed in each proportion. Concrete mechanical strength was tested, and the figures and tables show the average strength and standard deviations.

3.1 Strength properties of bacterial concrete

The addition of bacterial concrete may positively impact the mechanical properties of the concrete. The robust presence of many bacteria had a negative impact on the enhancement of strength properties, as the bacterium sample was significantly less strong than the reference mixture. The compressible modulus of a structure refers to its ability to withstand an axial load. The compression strength of the specimen was found to be 0.007 N/mm^2. As seen in Tables 2 and 3 [34], the durability of concrete has excellent tensile strength compared to ordinary concrete.

Table 2 Comparison of split tensile strength of conventional concrete and bacterial concrete

References	No. of days	Split tensile strength of conventional concrete cylinders, N/mm^2	Split tensile strength of bacterial concrete cylinders, N/mm^2	% increase in strength
[35]	28 days	3.78	4.23	13.12
[36]		4.62	4.44	14.12
[37]		4.82	5.03	14.23
[38]		4.88	5.12	14.63
[39]		4.92	5.22	15.69

The sample is submerged in 0.3 and 0.6 L of a 1-day-old stock cultured of B. Sphaericus for the primary treatment before being immersed in the nourishment liquid for 24 days because the bacterium inside the samples is the primary source of the ureolytic activities. Choosing a treatment depends on its economic viability in light of the many mechanisms shown in Table 3 [26].

Table 3 Comparison of compressive strength of conventional concrete and bacterial concrete

References	No. of days	Compressive strength of conventional concrete cubes, N/mm^2	Compressive strength of the bacterial concrete cylinders, N/mm^2	% increase in strength
[40]	28 days	19.25	18.22	30.22
[41]		24.66	19.52	32.21
[42]		28.66	20.33	35.66
[43]		33.25	23.33	40.22
[44]		43.66	34.22	46.12

An experiment using the RILEM 25 PEM (II-6) sorptivity test was used to measure the increase in the compressive water load. Water permeability suction was used to evaluate the bacterial concrete's ability to absorb water from regular concrete. When the value is less than 1, the initial moisture absorption is shown to decrease, and when the value is more significant than 1, the relative water absorption increases. As a result of the addition of calcite-precipitating bacteria, it can be done to create mortar with a self-healing capacity and increased strength through decreased sorptivity and adsorption capacity. As a result, the two bacteria, Bacillus megaterium, and Bacillus subtilis, were used because they met all of the demands. Using significant quantities of supplementary cementitious materials lowers material costs while requiring less energy and emitting less CO_2 than concrete. The use of metakaolin reduced the formation of tiny fractures and decreased. The relative initial surface absorption index suggested by [31] expressed the outcome. The experiment was performed on various specimens and discovered that ordinary concrete exhibits a lesser capillary index score.

3.2 Permeability test

The test involves subjecting a known-size mortar or concrete specimen in a customized cell to a known hydrostatic pressure from one side, determining the amount of water percolating through it over an agreed-upon interval, and calculating the value of the coefficient for permeability. Using the premise of the Hagen- Poiseuille correlation for laminar flow of compression and expansion through a porosity material possessing though only under stable equilibrium, the RILEM- CEMBUREAU approach was utilized to determine the flowability. Martin Sommer's air permeation study was employed to gauge the oxygenation flow velocity. It was discovered that the durability of concrete has less penetration than ordinary concrete [26]. The test involves subjecting a known-sized mortar or concretes the sample in a customized cell to a known hydrostatic pressure from one side, determining the amount of water flowing through it over an agreed-upon interval, and calculating the resulting coefficient of permeability. The test allows for measuring the water entering and leaving the specimen being tested. Water permeability is another critical element of the self-healing properties of concrete. Following the splitting test, the concrete specimen was fractured. A small amount of fluid immigrated into the fissures during the breaking test. After placing the specimen in the drying chamber to wait for the gel solution and create porous material, the cylinder was submerged in the water for three days. After three days, remove the cylinders and dry them.

The vacuum saturating used in the initial surface absorption test enables establishing a stable flow state in a specimen after it has been sucked in the high vacuum for two to three hours and contra waters have been introduced. For 24 hours, the cylinder was kept entirely submerged in water, and as a result of the fully immersed specimen, the pressure was lost. The samples must be elongated in shape and have a height equal to their diameter. The standard sample size is 150 mm in diameter (and height). In the case of samples with aggregates with nominal sizes less than 20 mm, the specimen's diameter (and height) may be decreased to 100 mm. In the case of samples containing particles than 40 mm in small size, the specimen's circumference (and height) must be at least four times the aggregate's nominal size. After that, the cylinder was removed to prepare for the initial surface absorption test. The entire setup was kept dry so the specimen remained saturated during the experiment. The time it took for the water table to drop from the top of the tube to the bottom throughout 30 days of evaluating the experiment with the fracture was monitored. Using Darcy's law, the continuity formula can compute the specimen's moisture absorption coefficients. Following the investigation, it was discovered that the optimal solution ranged from 4×10^{-6} m/s to 7×10^{-6} m/s, with the final price of k being 10^{-6} m/s, indicating that the silicone gel in the fracture had a restricted capacity to lessen the hydraulic conductivity. The crack's breadth started at 0.5 mm and shrank to 0.35 mm.

3.3 Stress-Strain Relationship of Concrete

Concrete's response under stress and pressure determines its hardness. The spherical specimen was examined using a universal testing machine with a 3000 kN capability, and the results are listed in Table 4 [18].

Table 4 Comparison of Stress-Strain behaviour of bacterial concrete and controlled concrete

References	Grade of concrete	Controlled concrete		Bacterial Concrete	
		Stress	Strain	Stress	Strain
[45]	M20	21.11	0.0007	20.11	0.0008
[46]	M25	27.56	0.0009	26.22	0.0008
[47]	M30	32.56	0.0011	31.46	0.0009
[48]	M60	61.25	0.0016	60.33	0.0010

4 Conclusions

Incorporating bacteria into concrete is quite useful since it increases the quality of the concrete, which is higher than conventional concrete. Bacteria fix concrete fractures through the production of calcium carbonate crystals which block the cracks and repair it. A number of researchers have conducted study on the self-healing characteristic of concrete and discovered that bacteria enhances the quality of conventional concrete by 13.12%, 14.12%, 14.23%, 14.63% and 15.69% in 28 days. The formation of calcium carbonate crystals reduces the water permeability by reducing fracture width from 0.5 mm to 0.35 mm. Compressive strength increased by 30.22%, 32.21%, 35.66%, 40.22%, and 46.12% in 28 days, and it was discovered in the mathematical model that bacterial concrete has a greater value of stress and strain than controlled concrete for the high strength grade of concrete. Based on the literatures, due to the use of self-healing material in the concrete, frequent inspection will be less necessary. Therefore, the potential benefits of self-healing concrete in terms of reducing maintenance costs, increasing service life, and enhancing sustainability make it an attractive option for the construction industry. However, further research and development are needed to overcome the existing challenges and optimize the performance of self-healing concrete.

References

1. K.V. Tittelboom, N. D. Belie, W. D. Muyncka, W. Verstraete, Use of bacteria to repair cracks in concrete, Cement and Concrete Research, **40 (1)**, 157-166 (2008). https://doi.org/10.1016/j.cemconres.2009.08.025.
2. L. Tan, B. Reeksting, V. Ferrandiz-Mas, A. Heath, S. Gebhard, K. Paine. Effect of carbonation on bacteria-based self-healing of cementitious composites. Construction and building materials, **257**, 119501 (2020). https://doi.org/10.1016/j.conbuildmat.2020.119501.
3. R. Siddique, N. K. Chahal, Effect of ureolytic bacteria on concrete properties, Construction and Building Materials, **25**, 3791–3801 (2011) https://doi.org/10.1016/j.conbuildmat.2011.04.010.
4. A. E. Enein Ali, F. Talkhan, A. Gawwad, Application of microbial cementation to improve the physical-mechanical properties of cement mortar, Housing and Building National Research Center, (2013).
5. H.M. Jonkers, A. Thijssen, O. Copuroglu, E. Schlangen, Application of bacteria as self-healing agent for the development of sustainable concrete, Proceedings of the 1st International Conference on BioGeoCivil Engineering, (23–25 June 2008), Delft, The Netherlands.
6. K. Santhosh, S.K. Ramachandran, V. Ramakrishnan, S.S. Bang, Remediation of concrete using microorganisms, American Concrete Institute Materials Journal, **98** (2001) 3–9. DOI: 10.14359/10154.
7. J.L. Day, V. Ramakrishnan, S.S. Bang, Microbiologically induced sealant for concrete crack remediation, 16th Engineering Mechanics Conference, (16–18 July 2003), Seattle, Washington.
8. S.S. Bang, J.K. Galinat, V. Ramakrishnan, Calcite precipitation induced by polyurethane- immobilized Bacillus pasteurii, Enzyme, and Microbial Technology, **28 (4),** (2001) 404–409. https://doi.org/10.1016/S0141-0229(00)00348-3.
9. J. Dick, W. De Windt, B. De Graef, H. Saveyn, P. Van der Meeren, N. De Belie, W. Verstraete, Bio-deposition of a calcium carbonate layer on degraded limestone by Bacillus species, Biodegradation, **17 (4),** (2006) 357–367. DOI: 10.1007/s10532-005-9006-x.
10. H. W. Reinhardt, M. Jooss, Permeability and self-healing of cracked concrete as a function of temperature and crack width. Cem. Concr. Res, **33**, 981–985 (2003). https://doi.org/10.1016/S0008-8846(02)01099-2.
11. I. Justo-Reinoso, B.J. Reeksting, C. Hamley-Bennett, A. Heath, S. Gebhard, K. Paine, Air-entraining admixtures as a protection method for bacterial spores in self-healing cementitious composites: healing evaluation of early and later-age cracks, Constr. Build. Mater, **327**, 126877 (2022). https://doi.org/10.1016/j.conbuildmat.2022.126877.
12. H. Chen, C. Qian, H. Huang, Self-healing cementitious materials based on bacteria and nutrients immobilized respectively, Constr. Build. Mater, **126,** 297–303 (2016). https://doi.org/10.1016/j.conbuildmat.2016.09.023.
13. J. Zhang, Y. Liu, T. Feng, M. Zhou, L. Zhao, A. Zhou, Z. Li, Immobilizing bacteria in expanded perlite for the crack self-healing in concrete, Constr. Build. Mater, **148,** 610–617 (2017). https://doi.org/10.1016/j.conbuildmat.2017.05.021.

14. M. Rauf, W. Khaliq, R.A. Khushnood, I. Ahmed, Comparative performance of different bacteria immobilized in natural fibers for self-healing in concrete, Constr. Build. Mater, **258,** 119578 (2020). doi.org/10.1016/j.conbuildmat.2020.119578.
15. M. Zamani, S. Nikafshar, A. Mousa, A. Behnia, Bacteria encapsulation using synthesized polyurea for self-healing of cement paste, Constr. Build. Mater, **249,** 118556 (2020). https://doi.org/10.1016/j.conbuildmat.2020.118556.
16. J. Wang, K. Van Tittelboom, N. De Belie, W. Verstraete, Use of silica gel or polyurethane immobilized bacteria for self-healing concrete, Constr. Build. Mater, **26 (1),** 532–540 (2012). https://doi.org/10.1016/j.conbuildmat.2011.06.054.
17. C. C. Gavimath, B. M. Mali, V. R. Hooli, J. D. Mallpur, A. B. Patil, D. P. Gaddi, C.R.Ternikar and B.E.ravishankera, Potential application of Bacteria to improve the strength of cement concrete, (2010).
18. C. M. Aldea, W. J. Song, J. S. Popovics, S. P. Shah, Extent of healing of cracked normal strength concrete. J. Mater. Civ. Eng, **12,** 92–96 (2000). https://doi.org/10.1061/(ASCE)0899-1561(2000)12:1(92).
19. C. Edvardsen, Water permeability and autogenous healing of cracks in concrete. ACI Mater. J, **96,** 448–454 (1999).
20. S. Jacobsen, E. J. Sellevold, Self-healing of high strength concrete after deterioration by freeze/thaw. Cem. Concr. Res, **26,** 55–62 (1995). https://doi.org/10.1016/0008-8846(95)00179-4.
21. V. Wiktor, and H. M. Jonkers, Quantification of crack-healing in novel bacteria-based self-healing concrete, Cement and Concrete Composites, **33 (7),** 763-770 (2011). https://doi.org/10.1016/j.cemconcomp.2011.03.012.
22. C. A. Clear, The Effects of Autogenous Healing upon the Leakage of Water through Cracks in Concrete; Cement and Concrete Association: Slough, UK, p. 28 (1985).
23. M. Şahmaran, S. B. Keskin, G. Ozerkan, I. O. Yaman, Self-healing of mechanically-loaded self-consolidating concretes with high volumes of fly ash. Cem. Concr. Compos, **30**, 872–879 (2008). https://doi.org/10.1016/j.conbuildmat.2020.118556.
24. S. Reddy, Rajaratnam, M. SeshagiriRao, Sasikala. Mathematical Model for Predicting Stress-Strain Behaviour of Bacterial Concrete, **8 (7),** 470-479 (2010).
25. J. Dick, W. Windt, B. Graef, H. Saveyn, P. Meeren, N. De Belie, W. Verstraete, Biodeposition of a calcium carbonate layer on degraded limestone by Bacillus species, Biodegradation, **17 (4),** 357–367 (2006).
26. F. Hammes, N. Boon, J. de Villiers, W. Verstraete, S.D. Siciliano, Strain-specific ureolytic microbial calcium carbonate precipitation, Applied and Environment Microbiology, **69 (8),** 4901–4909 (2003).
27. Office of Environmental Health and Safety; Washington State Department of Health. Larvicide: *Bacillus sphaericus*, (2006). <http://www.doh.wa.gov/ehp/ts/Zoo/WNV/larvicides/Bsphaericus.html#2>.
28. S. Sookie, S. Bang, K. Johnna, K. Galinata, V. Ramakrishnan, Calcite precipitation induced by polyurethane-immobilized Bacillus, (2009).
29. S.J. Park, Y.M. Park, W.Y. Chun, W.J. Kim, S.Y. Ghim, Calcite-forming bacteria for compressive strength improvement in mortar, J. Microbiol. Biotechnol, **20 (4)** (2010) 782–788. DOI:10.4014/jmb.0911.11015.
30. R. Peruzzi, T. Poli, L. Toniolo, The experimental test for the evaluation of protective treatments: a critical survey of the capillary absorption index. J Cult Herit, **4(3),** 251–4 (2003). https://doi.org/10.1016/S1296-2074(03)00050-5.

31. A.K. Galwey, M.E. Brown, Thermal decomposition of ionic solids, Studies in Physical and Theoretical Chemistry 86 (1999).
32. H. Jonkers, Bacteria-based self-healing concrete, HERON **56 (1),** 1-12 (2011).
33. R. Peruzzi, T. Poli, L. Toniolo. The experimental test for the evaluation of protective treatments: a critical survey of the capillary absorption index. J Cult Herit, **4(3),** 251–4 (2003).
34. E.V. Tararushkin, V.V. Pisarev, A.G. Kalinichev, Atomistic simulations of ettringite and its aqueous interfaces: structure and properties revisited with the modified ClayFF force field, Cem. Concr. Res, **156,** 106759 (2022).
35. R. Dupuis, R.J.M. Pellenq, Alkali silica reaction: A view from the nanoscale, Cem. Concr. Res, **152,** 106652 (2022). https://doi.org/10.1016/j.cemconres.2009.08.025
36. T. Honorio, M. Maaroufi, S. A. Dandachli, A. Bourdot, Ettringite hysteresis under sorption from molecular simulations, Cem. Concr. Res, **150,** 106587 (2021).
37. F. Lolli, H. Manzano, J.L. Provis, M.C. Bignozzi, E. Masoero, Atomistic simulations of geopolymer models: the impact of the disorder on structure and mechanics, ACS Appl. Mater. Interfaces, **10 (26),** 22809–22820 (2018).
38. A. Kunhi Mohamed, S.A. Weckwerth, R.K. Mishra, H. Heinz, R.J. Flatt, Molecular modeling of chemical admixtures; opportunities and challenges, Cem. Concr. Res, **156,** 106783 (2022). https://doi.org/10.1016/S0008-8846(02)01099-2.
39. R. J. M. Pellenq, A. Kushima, R. Shahsavari, K. J. Van Vliet, M. J. Buehler, S. Yip, F. J. Ulm, A realistic molecular model of cement hydrates. Proceedings of the National Academy of Sciences, **106(38),** 16102-16107 (2009).
40. Alex, A. and Masoero, E., Autogenous healing in cement: A kinetic Monte Carlo simulation of $CaCO_3$ precipitation. In Computational Modelling of Concrete and Concrete Structures, 102-106, (2022).
41. CRC Press. Lee, J., Patel, D.S., Kucharska, I., Tamm, L.K. and Im, W., Refinement of OprH-LPS interactions by molecular simulations. Biophysical journal, **112**(2), pp.346-355 (2017)
42. Jayathilake PG, Gupta P, Li B, Madsen C, Oyebamiji O, González-Cabaleiro R, Rushton S, Bridgens B, Swailes D, Allen B, McGough AS. A mechanistic Individual-based Model of microbial communities. PloS one. **12(8)**:e0181965, (2017).
43. C. Z. Qin, S. M. Hassanizadeh, A. Ebigbo, Pore-scale network modeling of microbially induced calcium carbonate precipitation: Insight into scale dependence of biogeochemical reaction rates. Water Resources Research, **52(11)**, 8794-810 (2016).
44. D. Redondo, Eduardo, A. Patrick, Bonnaud, H. Manzano, A comprehensive review of CSH empirical and computational models, their applications, and practical aspects, Cement and Concrete Research, **156**, 106784 (2022). https://doi.org/10.1016/S0008-8846(02)01099-2.
45. M. Dhaarani, K. Prakash, Durability study on hvfa based bacterial concrete—a literature study, International Journal Structural & Civil Engineering, ISSN: 2319-6009 (2014).
46. R. Peruzzi, T. Poli, L. Toniolo, The experimental test for the evaluation of protective treatments: a critical survey of the "capillary absorption index". Journal of Cultural Heritage, **4(3)**, 251-4 (2003).

47. S. Ichiigarashi, A. Hosoda, Takashihitomi, K. Ichiimamoto, Technical committee on self-healing / Repairing Technology in cement-based materials.
48. Wang, J., Van Tittelboom, K., De Belie, N. and Verstraete, W., Use of silica gel or polyurethane immobilized bacteria for self-healing concrete. Construction and building materials, **26(1)**, 532-540 (2012).

Experimental Investigation on Geopolymer Concrete:- Sustainable and Eco-Friendly Construction Material

Amina Rafeek[1], ***Sundar* Kumar.S.**[2] **and *Indu* Susan Raj**[3]
*1 PG Student, ,Mar Athanasius College of Engineering, Kothamangalam, Kerala-686691;*Email: aminarafq1999@gmail.com.
[2] Principal Scientist, AML Lab, CSIR-SERC, Taramani,Chennai-600113; Email: ssk@serc.res.in
*3 Assistant Professor, Mar Athanasius College of Engineering, Kothamangalam, Kerala-686691;*Email: indususanraj@mace.ac.in

Abstract

The research for the new environmental friendly construction material that will match the durability of ancient concrete has provoked interest into the study of alkali activated cementitious systems over the last two decades. It results in the development of Geopolymer Concrete. This paper represents the development of geopolymer mixes of various ratios of fly ash and GGBS and their mechanical properties were studied. aa Literature review on mechanical and durability properties of geopolymer concrete also been done. The ratio of sodium hydroxide to sodium silicate is 1:2 which is kept a constant to prepare the mix and alkaline liquid to binder ratio as 0.70, but changing the concentration of sodium hydroxide solution as 6M, 4M and 3M to prepare the mix Id. i.e. F75G25 , F50G50 and F25G75, respectively (where F and G are Fly Ash and GGBS and the numerical value indicates the percentage of replacement of fly ash by GGBS) were prepared in AML laboratory ,CSIR-SERC, Chennai. Ambient curing of concrete at room temperature was adopted.The compressive strength at age of 3, 7 and 28 days were determined. The strength of geopolymer concrete was increased with increase in percentage of GGBS in a mix. It was observed that the mix Id F25G75 gave maximum compressive strength for concentration of NaOH as 3M . Leaching of GPC reduces with increase in percentage of GGBS in a mix. Thus the geopolymer concrete have a relatively higher strength.

Keywords: Geopolymer concrete, Compressive strength, mix, Alkaline solution,Ambient curing, Groundgranulated blast furnace slag, Flyash.

1. Introduction

Concrete is the most commonly used construction material; it's usage by communities across the globe is second only to water. Ever grander building and infrastructure projects require prodigious quantities of concrete with its binder of Portland cement whose manufacture, as we have seen, is accompanied by large emission of carbon dioxide. Ranging from 0.66 to 0.82kg of CO_2 emitted for every kilogram manufactured. The contribution of the production of OPC is approximately 5-7% of global anthropogenic CO_2emissions. The key cases of high CO_2 emissions arising from OPC manufacture have been attributed to i) Calcinations of lime stone, one of the key ingredients, which leads to formation and release of CO_2 and ii) High energy consumption during manufacturing, including heating raw materials within a rotating kiln at temperatures greater than 14000°C.
The call for the day is, "*Sustainable Development*" which demands the new concrete technology which uses less natural resource, energy and generates less CO_2 without compromising on strength and durability aspects. This burgeoning worldwide demand for concrete is a great opportunity for the development of geopolymer concrete and geopolymer cements of all types, with their much lower tally of CO_2. As of now, waste materials have partly been used as aggregates and fillers in concrete. The research for the new environmental friendly construction material that will match the durability of ancient concrete has provoked interest into the study of

© The Authors, published by EDP Sciences. This is an open access article distributed under the terms of the Creative Commons Attribution License 4.0 (https://creativecommons.org/licenses/by/4.0/).

alkali activated cementitious systems over the last two decades. One effort to combat short fall is the development of alternate binders to Portland cement aiming at to reduce the environmental impact of construction, use of greater proportion of waste pozzolan, and also due to improve concrete performance.

Geopolymer is an ecofriendly material produced as a by-product from industrial wastes such as fly ash, ground granulated blast slag, metakaolin, rice husk etc. The use of geopolymer reduces the carbon footprint. This reduces the demand on conventional cements. Geopolymers have vast advantages over conventional cement in view with mechanical and durability properties. Commercially geopolymers are used for fire and heat resistant and as new binders for fire-resistant fiber composite toxics and at radioactive waste encapsulation and for producing an alternative to conventional cements.

Geopolymers are more resistant to corrosion and fire and has high compressive and tensile strength. They gain full strength faster and has less shrinkage than the conventional concrete. Geopolymer concrete is a viable alternate to Portland cement concrete. The process of geopolymerisation has summarized in 3 steps such as dissolution of silica and alumina from the source material through the action of hydroxide ions,transportation or orientation or condensation of precursor ions into monomers.It essentially consists of covalently bonded (Si-O-Al-O) sialiate monomers. It is an inorganic by product material which is rich in silicon (Si) and alumina (Al),that react with alkali activators to from 3D polymeric chains of saliate and poly(sialiate).

Zhang and Wang(2013) studied the phenomenon of efflorescence of fly-ash geopolymers is mainly due to sodium carbonate heptahydrate ($Na_2CO_{37}H_2O$). The efflorescence has occurred in 3hrs in specimens which was in contact with water at the bottom and it is faster in 25°C sealed aged geopolymers. Therefore, hydrothermal curing reduces the formation of efflorescence. The potential of efflorescence is compared by measuring cation concentrations, pH value and electrical conductivity of geopolymer leaching solutions. The use of soluble silicate present activator can achieve high strength as well as early age efflorescence under wet conditions and restricts long term efflorescence potential. The pore solution analysis result shows that if soluble silicate content increases, then the geopolymer mixture tends to decrease the alkali metal concentrations. And compressive strength test for FA geopolymer concrete with 100% sodium silicate activator, FA geopolymer concrete with 100% NaOH solution as activator and geopolymer concrete of 80% FA and 20% slag are cured at 80°C and 25°C. The results show that geopolymer concrete containing slag's strength has increased eloquently at 25°C compared to other mix. And the geopolymers cured at 80°C, FA geopolymer concrete with 100% sodium silicate activator has increased strength compared other geopolymer concrete. The addition of slag reduces the alkali leaching and hence it reduces the leaching rate at initial stage only. From electrical conductivity it is found that Na^+, K^+, OH^- ions are the main contributors to leaching rate.Madheswaran and Gnanasundar(2013) has done an experimental investigation on use of three different molarities of NaOH range 3M, 5M and 7M and their influence on three different types of mixes in combination with FA and GGBS to study the effect of AAS in geopolymer concrete. The mixes are 50% GGBS and 50% FA, 75% GGBS and 25% FA and 100% GGBS. The result of compressive strength test of cube size 150 x 150 x 150mm shows that strength increases with the molarity of NaOH. For the mix of 100% GGBS and 7M NaOH, it gives maximum compressive strength of 60MPa in 28 days. When the addition of GGBS increases then the compressive strength is also increased. From the split tensile result of 150mmϕ 300mm height cylinders it has been observed that when the quantity of GGBS increase the split tensile strength also increases.Chamila Gunasekara(2016) has studied the durability characteristics of four different flyash geopolymer concretes and their permeation properties was carried out for up to an age of 1 year. The flyash's were taken from four different power plants in Australia. The AAS is a mixture of sodium silicate and sodium hydroxide solution (15M). The aggregates used are river sand as fine aggregate and 7mm & 10mm size coarse aggregates were used. The flyash geopolymer concrete has slightly lower density than plain cement concrete and a collapsed type of slump. In these four types of flyash, gladstone geopolymer has highest compressive strength of about 48.7MPa and Collie geopolymer has lowest value of 23.5MPa in 28days. From the compressive test they've found that there is an-going geopolymerization as the compressive strength is increased with time in flyash geopolymer concrete. Gladstone, Pt. augusta and Tarong geopolymers achieved a compressive strength more than 30MPa in 3days under exposure classification B1, B2. Gladstone geopolymer concrete has enough 50MPa compressive strength of maritime structure in sea tidal zones under exposure classification C1, C2. From SEM data it has been observed that there is an increase in the aluminosilicate gel in the concrete matrix and a decrease in the quantity of micro cracks for all geopolymers. From UPV test it has been found that only gladstone geopolymer concrete has categorised being in good condition. Gladstone and Pt. Augusta shows reduction in water absorption with time and are classified as low permeable concrete and as well as low air permeability index with time. But Tarong is significantly performing well with respect to water absorption, AVPV, strength test, UPV, permeability data's observed over a period of 1year. A salt ponding test has been done for a term of 90days to observe its resistance to corrosion. Gladstone and Pt. Augusta geopolymer concrete showed high surface chloride concentration at early age which was higher than plain cement concrete and blended cement concrete but the surface chloride concentration of all four geopolymer concretes reduced with age of concrete and varied between 0.34 to 0.39% chloride by weight of concrete at 1 year. The chloride diffusion coefficient of all four flyash geopolymer were between 15.8 x 10^{-11} and 21.5 x 10^{-11} m^2/s at 28days but these were decreased with age of concrete. Hence an improvement in the durability properties of all four flyash geopolymer concrete with time. Zengqing Sun(2020) three types of geopolymers were tested for leaching, they are flyash geopolymer (FAG), metakaolin geopolymer (MKG) and commercial geopolymers (CG) are used. The horizontal dynamic surface leaching of Na, K, Al, Si, Ca, Ba, As, Cd, Co, Cr, Cu, Mo, Ni, Pb, Sb, Se, Tl, V, Zn were studied and was done for samples that are cured for 56days before leaching. FAG has lower reaction and the unreacted alkali in FAG has higher pH value and releases more Na because of using sodium-based activators. Si is more easily leached than Al in FAG and CG. The results show there's strength loss after leaching in deionized water of FAG, MKG and CG samples of about 11.8%, 9.4% and 14.9%. The 56day compressive strength of MKG is 68MPa, which more compared to FAG and CG.Rachit ghosh(2018) investigated leaching on three flyash collected from Tata Power Jojobera Plant, Rawan Plant and Ennore Plant and named as FA1, FA2 and FA3 resp. GGBS were collected from Tata Steel and bottom ash (BA) from Tata power plant. The flyash based geopolymer concrete is a mixture of flyash, GGBS, bottom ash as fine aggregate and crushed stone aggregate. The AAS used were NAOH and NA_2SiO_3 (1:1) as alkali-1 (6M), alkali-2 (8M) and alkali-3 (10M). From XRD it is observed the flyash's from different plants are quartz, mullite, hematite and magnetic. In Toxicity characteristic leaching procedure (TCLP), 1M sodium acetate buffer was used as an extraction liquid at pH 4.99 for TCLP and for batch leach test the pH value is 7.2. Mn, Cu and Mg were leached out at higher amount in all Flyash samples and Ni, Fe, Pb were leached out at a lower concentration. It is also found that metals concentration in TCLP was higher than batch leach test since the solubility of metals decreases with increasing pH value, because of insoluble hydroxides at high pH values. From mercury intrusion, the pores are uniformly distributed and the variation in number of pores is less in 6MFA3GC when compared to 6MFA1GC and 6MFA2GC. From the compressive strength test, the 6MFA3GC reported the maximum strength which is then followed by 6MFA2GC and 6MFA2GC but when compared with different concentration of alkali, the maximum strength is achieved by 6MFA3GC and minimum strength by 10MFA1GC. This is because the 1 day to 7days strength was slow due to alkali's reaction but it was totally reverse after 28day strength. The leaching out and efflorescence behaviour and its effect were studied for up to 180 days, it has been found that leachates of heavy metals and trace elements were below the permissible limits as per IS code. And also there was no visual deterioration of geopolymer concrete specimens when exposed to weathering condition up to 180 days.

2. Materials and Methods

The materials used in the study are class F fly ash (FA) and ground granulated blast slag (GGBS). The concentration sodium hydroxide (SH) solution in this study was 6Molar, 4 Molar and 3 Molar. The ratio of sodium hydroxide to sodium silicate (SS) is maintained at 1:2. The specific gravity and bulk density of low calcium class F fly ash was 2.8 and 995 kg/m3 respectively. The specific gravity and bulk density of GGBS was 2.9 and 1250kg/m^3 with a fineness of 400 m^2/kg.
The oxide composition of fly ash and GGBS was obtained from XRF analysis is given in table 1.Sodium hydroxide was used in the form of flakes and sodium silicate in liquid form. M sand is used as fine aggregate and coarse aggregate of 10mm and 20mm down size is used in 60% and 40% of total weight.

Table 1 Oxide Composition of Fly Ash and GGBS

Materials	SiO_2	Al_2O_3	Fe_2O_3	CaO	MgO	Na_2O_3	K_2O	TiO_2	Mn_2O_3	SO_3
Fly Ash	62.1	27.44	4.57	0.83	0.55	0.04	1.17	1.09	0.04	0.04
GGBS	43.4	12.5	-	40.3	1.5	0.9	0.6	-	-	-

While preparing Alkali activator solution (AAS), first the SH flakes were dissolved a day before casting due to rise in temperature during addition of flakes in water then SS solution was added 2mins before casting of concrete. Slump value was found for all the mixes.

Table 2 Composition and Properties of Sodium Silicate

Na_2O (%)	SiO_2 (%)	Total Solids (%)	Viscosity (Ns/m^2)	Specific gravity
14.2	31.2	45.4	900	1.4

The specific gravity and bulk density was 2.65 and 1550kg/m^3 respectively for M sand and for coarse aggregate its 2.71 and 1600kg/m^3. The ratio of Fly ash to GGBS in this study is 75:25, 50:50 and 25:75, three mixes were developed for this ratio for 6M, 4M and 3M respectively. FA, GGBS, fine and coarse aggregates were mixed dry in 50Kg mixer pan for few minutes the AAS is added little by little looking at the consistency of the concrete. The addition AAS can lesser or more than the calculated value. The geopolymer concrete cubes of all the mixes were cured under ambient temperature. The details of the 3 mixes are given in the table 3 .

Table 3 Quantity of Mixes

Mixes		GGBS	Fine Aggregate	Coarse Aggregate		Activators	
				10mm	20mm	AAS	Molarity
GM1	300	100	655	606	404	260	6
GM2	200	200	645	594	396	260	4
GM3	100	300	640	588	392	260	3

The slump test is shown in Fig 1 and its values for various mixes are depicted in table 4.

Figure 1 Slump test

Table 4 Slump Value

Mix ID	Slump Value (mm)
GM1	195
GM2	198
GM3	190

2.1 Tests on Concrete

2.1.1 Compressive strength of concrete

The compressive strength of concrete was done to determine the strength of 100mm cube on 3rd day, 7th day and 28th day in UTM (Universal Testing Machine) shown in Fig 2. The compressive strength results are depicted in table 5.

Figure 2 Compression Test

Table 5 Compressive strength results

MIX ID	3rd days strength, MPa	7th day strength, MPa	28th day strength, MPa
GM1	8.93	15.02	28.48
GM2	14.56	29.68	46.25
GM3	28.23	38.12	53.5

3. Results and Discussions

The detailed analysis has been carried out for,and the results are described in the following.

- A total of three geopolymer mixes were studied.
- The geopolymer mixes achieved a very good workability in terms of slump values. The mixes also had sufficient workability retention time. Though slump retention was not measure, the mixes started hardening only after 30-40 minutes after initial mixing.
- Comparative Compressive Strength Graph is shown in Fig3. The highest strength was achieved by the Mix GM3.
- All the geopolymer concrete mixes had a faster rate of strength gain when compared to the rate normally one would anticipate from Portland cement concrete.
- Initial stages leaching was observed, but it decreases with age and Leaching of GPC reduces with increase in percentage of GGBS in a mix.
- Hence geopolymer concrete of nominal strength can be produced with relative mild activators. These mixes can be handled easily, produced in situ and in large quantities with existing infrastructure itself.

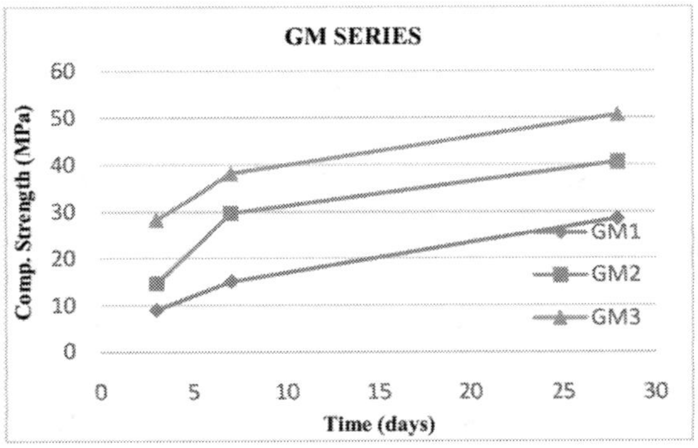

Figure 3 Comparative Compressive Strength Graph

References

1) Ghosh, R., Gupta, S. K., Kumar, A., & Kumar, S. (2018). Leaching and efflorescence effects in geopolymer concrete. Journal of Metallurgy and Materials Science, 60(2), 79-88.
2) Gunasekara, C., Law, D. W., &Setunge, S. (2016). Long term permeation properties of different fly ash geopolymer concretes. Construction and Building Materials, 124, 352-362. https://doi.org/10.1016/j.conbuildmat.2016.07.12
3) Madheswaran, C. K., Gnanasundar, G., &Gopalakrishnan, N. (2013). Effect of molarity in geopolymer concrete. International Journal of Civil & Structural Engineering, 4(2), 106-115.

4) Sun, Z., &Vollpracht, A. (2020). Leaching of monolithic geopolymer mortars. Cement and Concrete Research, 136, 10616
5) Z. Zhang, H. Wang, J.L. Provis, & A. Reid, (2013). Efflorescence: a critical challenge for geopolymer applications In Concrete Institute of Australia's Biennial National Conference 2013 (pp. 1- 10) . Concrete Institute of Australia.

Generation of Facemask in India during the COVID - 19 Pandemic

Krithiga Palanisamy[1]*, *Riyaz Ahil* Nizamudheen[2], *Sabari Balaji* Sivakumar[3]

[1]Assistant professor, Department of Civil Engineering, Kongu Engineering College, Perundurai, Erode, Tamil Nadu, India.
[2,3]Pre Final year students, Department of Civil Engineering, Kongu Engineering College, Perundurai, Erode, Tamil Nadu, India.

Abstract. The continuing COVID-19 outbreak has urged enormous using up of Personal Protective Equipment (PPE). Individual usage of the facemask become most popular to prevent the spread of the virus. Wearing facemask becomes necessary and part of the various regulations an guidelines encouraged by the government across the country worldwide. The increase usage of the facemask creates threat to the environment that leads to the pathway of the pollution. This article is the first to estimate the daily usage of the facemask generation in India. Separate state-wise usage of the facemask were highlighted. The results clearly portraits the usage of facemask consumption increase year by year. Finally, the outlook and challenges on additional development and applications of the facemask were discussed.

Keywords: Facemask, COVID-19, pandemic, infectious waste, disposal

1. Introduction

The unexpected breakout of the covid-19 widespread on March 2020 directed to disaster all over the world. World Health Organization WHO, 2020 that includes home quarantine rules, has published many measures and safety precautions and mandatory usage of PPEs that includes cloth mask, face shield, gloves [1]. China reports the first cases to the WHO in December 2019. Since, 224 countries and territories have reported 551.3 million Covid cases and 6.3 million death [2]. The continent such as Europe and Asia adds to about 206,490,180 and 162,232,180. The countries like North America leads to 106,318,690, South America to 60193747 followed by Africa (12,371,984) and Oceania (10,0983,52) [3]. According to July 2022, report says by Ministry of Health and Family Welfare, Government of India, active cases was set to be 119,457 [4]. The pandemic spread across India led to over 43,568,884 coronavirus cases whereas death rate rises to 525,270 report by Worldometer (as of July 2022) [5]. In India the first wave starts from Feb 15 2020, number of infected people said to be risen. Again, from May 2021 the second wave starts and number of cases (more than 3 million) is peak when compared to first wave. Among 28 states in India as of now, the Maharashtra have come up more than (7,972,474) confirmed covid-19 cases. The states such

*Corresponding Author: krithipalanisamy@gmail.com

© The Authors, published by EDP Sciences. This is an open access article distributed under the terms of the Creative Commons Attribution License 4.0 (https://creativecommons.org/licenses/by/4.0/).

as Kerala (6,634,722), Karnataka (3968365) followed by Tamil Nadu (3,473,116), Andhra Pradesh (2,321,379) and Uttar Pradesh (2,090,050) [6].

Currently, India is facing with different issues in waste management of the bulk of waste generation. During COVID-19, the generation of the biomedical waste in some regions reaches peaks. Central Pollution Control Board (CPCB, 2020) has framed several new guidelines, treatment facilities and disposal techniques [7]. Normally hospitals release some quantities of waste especially biomedical waste which some toxins that can disturb the environment [8]. Over two years, the outbreak of COVID-19 leads to way for the high generation of the hospital waste. Biomedical Waste is the waste produced during diagnosis, treatment or immunization of human or animal research activities [9]. HW contains more chemical properties like biochemical oxygen demand (BOD), Chemical oxygen demand (COD), nitrites, phosphorus and total solids [10]. As per Biomedical waste Management Rules 2016, the treatment of biomedical waste is essential to be treated and disposed through common Biomedical Waste Treatment or else in deep burial [11]. As per (CPCB, Delhi) moderately 3.54 TPD quantity of COVID-19 biomedical waste generation occurs during May 2022 [12]. Figure.1 shows the generation of COVID-19 related biomedical waste in State/UT. The irregular move of COVID-19 waste has bewildered the BMW quantity assessment and their organization [13]. Some studies in India shows the segregation and treatment methods of BWMs in detailed manner [14].

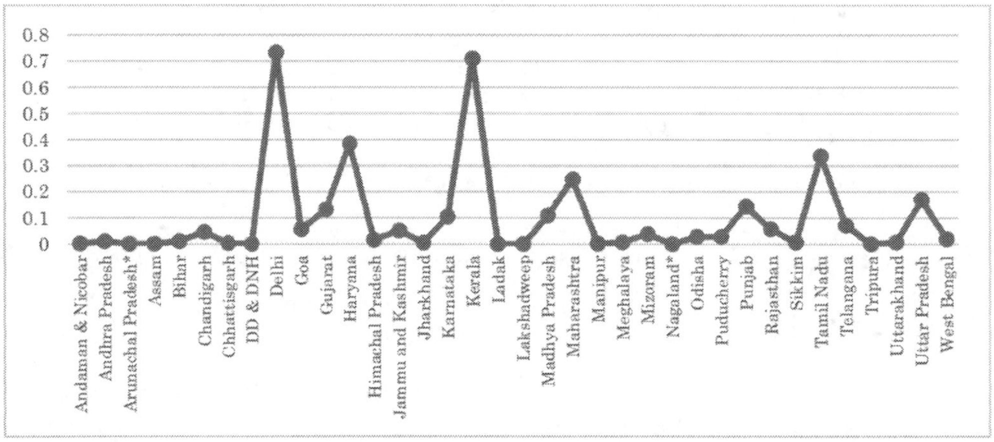

Figure 1: Details of COVID-19 state-wise BW generation (upto to May 2022)

During the pandemic period, the sudden increase of biomedical waste and usage of facemask tends to aerial. Currently world is facing large single use plastic demand in this pandemic period [15]. Many countries have their own guidelines based on the socioeconomic condition and healthcare specialists recommends the workers to wear the facemask, face shield and gowns [16]. Nowadays facemask becomes more widespread and discarded PPE. While the effectiveness of the facemask during the spread of corona viruses becomes uncertain, even though some people are also discriminating against those who are not wear mask. Anyhow, the usage of facemask is mandatory and becomes people practices. Various countries poses the people to use the single-use PPE in public places that create numerous waste management problems and could severely make worse further plastic pollution [17]. The throwing away of facemask could able to leach chemical contaminants and microfibers to the environment [18]. Amid, the usage of sudden increase in PPE, many studies have estimated the consumption of face masks in their concern area. Some urban population like Victoria estimated the usage of the facemask during the 1st and 2nd lockdown was said to be 5351520 per day respectively [19]. In other countries like India, the daily facemask was estimated to be 381,179,657 per day. In other hand, highly populated area like China was likely to be 989,103,299 day^{-1}. Totally, estimated daily facemask use in Asia was generated to be 2228170832 day^{-1} [20]. With this study, we conducted a case study in India, which have some improper waste management plans, based on the generation of facemask in each state under different circumstances.

2. Material and Methods

2.1 Study Area

India is located at 20.5937°N latitude and 78.9629°E longitude was selected for the study area. It is one of the vast geographical expanse and in the North is confined by Himalayas, the Arabian Sea in the West, Bay of Bengal in the east and Indian Ocean in the South. The second populous country and most democracy in the world. Area of about 3.28 million sq.km. Figure 2. Show the location of the study area.

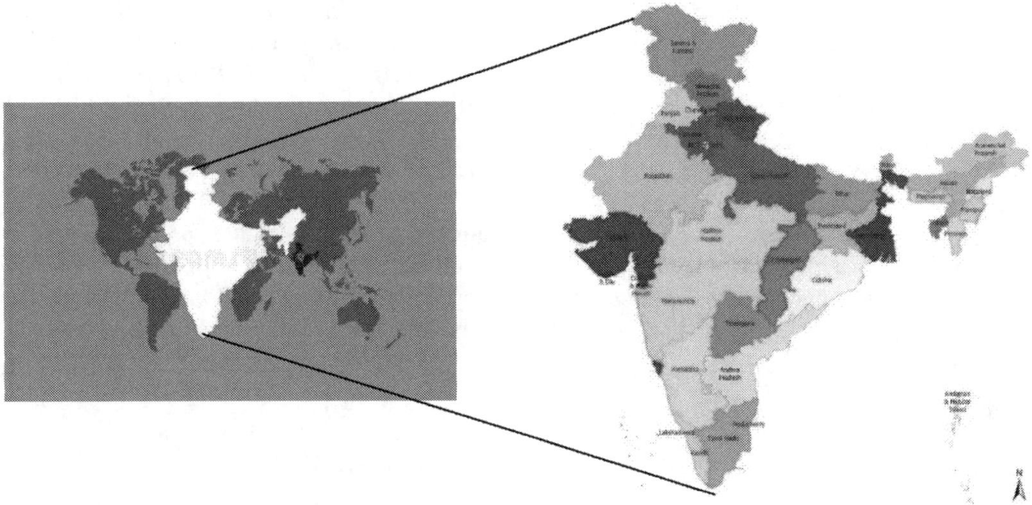

Figure 2. Location of Study Area

2.1.1 Population forecast

The population data for 2022 were reclaimed from India census data. The method was used to identify the spatial variation of the outbreak within the country. More importantly, this method will not fit with different countries across the world. These data's and information are essential to estimate the COVID-19 related facemask in India.

2.1.2 Estimation of daily facemask consumption

The consumption of the facemask is estimated by using the equation, which were taken from Nzediegwu and Chang [21].

$$D_{FM} = P \times Up \times F_{MAR} \times (F_{MGP}/10000)$$

Where

D_{FM} = Daily face mask use (pieces)
P = Population (persons)
U_P = Urban population (percentage)
F_{MAR} = Face masks acceptance rate – 80% [21].
F_{MGP} = Assumption that each person in the general population uses two facemask each day

The study is limited to provide a rapid estimation of potential facemask that is associated with the COVID-19 pandemic in Asia.

3. Result

This study tends to estimate the generation of facemask used in the India during COVID-19 pandemic period. The result shows that more people were infected in the concern states. The overall daily facemask waste generation for the entire India reaches 67,19,831 facemask per day under single use scenario. In general, the particular states like Chhattisgarh (119732) Uttar Pradesh (824890), Maharashtra (903706), Tamil Nadu (648156), follows by West Bengal (514493) seems to be high usage of facemask that was shown in Table 1. The other states usage was depicted in figure 3. This research used mathematical calculation to found out the generation facemask within India. Subsequently, it was predicted for estimating the usage of facemask depends on population strength, urban population, acceptance rate and the facemask daily usage. Finally, the equation was applied to estimate the facemask with the predicated population during the COVID 19 pandemic period for the healthcare waste management.

Table 1: Estimated daily facemask in India with confirmed COVID-cases

States	Predicted Population 2022*	Covid Cases**	Urban Population (%)	Acceptance Rate %	No. of Facemasks used per day	Total Facemasks
Andaman & Nicobar Islands	399,001	10,340	37.69%	80	2	2407
Andhra Pradesh	91,702,478	2,327,662	33.36%	80	2	489472
Arunachal Pradesh	1,711,947	65,034	22.94%	80	2	6284
Assam	35,998,752	731,662	13.99%	80	2	80637
Bihar	128,500,364	839,591	11.29%	80	2	232124
Chandigarh	1,158,040	95,046	97.25%	80	2	18020

Chhattisgarh	32,199,722	1,159,921	23.24%	80	2	119732
Dadra and Nagar Haveli	453,008	11,512	46.71%	80	2	3387
Daman and Diu	320,989		75.16%	80	2	2861
Delhi	19,301,096	1,944,978	97.5%	80	2	301098
Goa	1,521,992	250,862	62.16%	80	2	15140
Gujarat	70,400,153	1,244,114	42.59%	80	2	479847
Haryana	28,900,667	1,023,257	34.87%	80	2	161288
Himachal Pradesh	7,503,010	291,086	10.03%	80	2	12040
Jammu and Kashmir	14,999,397	457,517	27.38%	80	2	65709
Jharkhand	40,100,376	438,106	24.04%	80	2	154307
Karnataka	69,599,762	6,697,013	38.67%	80	2	430627
Kerala	34,698,876	6,697,013	47.70%	80	2	264822
Lakshadweep	66,001	11,412	78.07%	80	2	825
Maharashtra	124,904,071	8,022,781	45.22%	80	2	903706
Manipur	3,436,948	138,123	29.21%	80	2	16063
Meghalaya	3,772,103	94,409	20.06%	80	2	12113
Mizoram	1,308,967	231,066	52.11%	80	2	10914
Nagaland	2,073,074	35,628	28.85%	80	2	9573
Orissa	47,099,270	1,300,873	16.68%	80	2	125774
Punjab	30,501,026	768,502	37.48%	80	2	182909
Rajasthan	79,502,477	1,291,293	24.87%	80	2	316357
Sikkim	658,019	40,235	25.14%	80	2	2648
Tamil Nadu	83,697,770	3,522,142	48.40%	80	2	648156
Telangana	38,157,311	810,976	38.90%	80	2	237492
Tripura	4,184,959	103,270	26.17%	80	2	17524
Uttar Pradesh	231,502,578	2,097,156	22.26%	80	2	824890
Uttarakhand	11,700,099	440,063	30.22%	80	2	56592
West Bengal	100,896,618	2,074,550	31.86%	80	2	514493

*Data as on: https://www.indiacensus.net/

** https://www.mohfw.gov.in/

*** Data as on: https://www.worldometers.info

4. Discussion

In the year 2020, the usage of the facemask was found to be 381,179,657 [22]. In our study, the total generation of the facemask estimated to be 67,19,831 was seems to be high when compared to prior year. The research conducted by [21] figured out that the total number of facemask used daily was surged during COVID-19 pandemic.

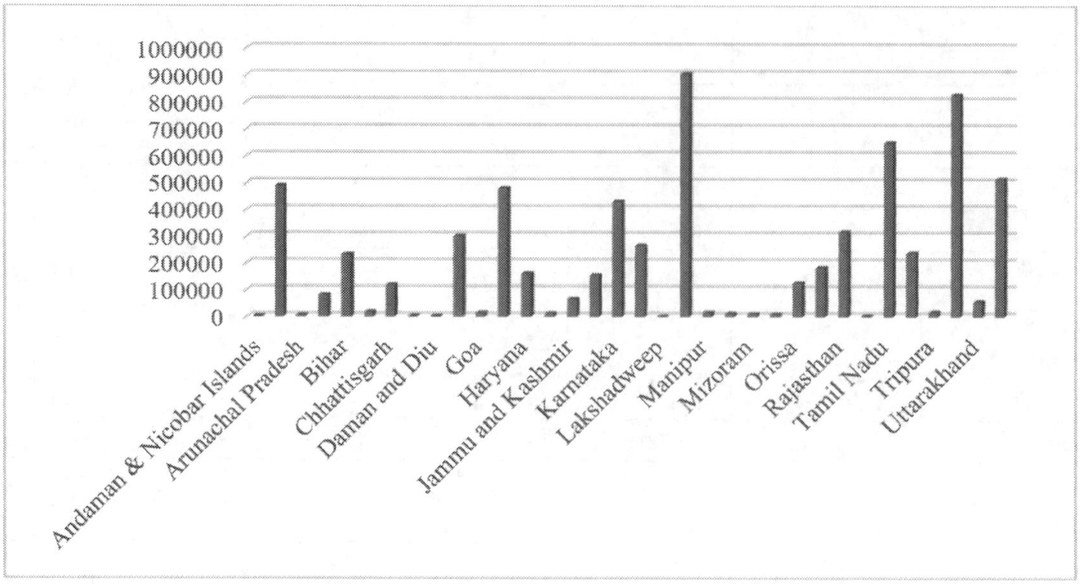

Figure 3: Facemask usage in India

This outbreak of the COVID-19 has caused not only the global health crisis but also its changed the waste generation patter [23]. The COVID-19 occurrence has increased the facemask in India and in other countries. The result of this study showed that also it lead to the huge height of facemask generation those reflects in municipal waste management, which adversely affect the environmental issues [24]. The daily utilization of the facemask further reduces the spread of the COVID-19 outbreak has resulted in the widen part facemask waste. In other way, if the waste is not properly managed it leads to spread of more diseases. Improper handling of these waste create the formation of plastic pollution [25]. Due to increase of COIVD cases year by year, the usage of facemask and generation of medical waste reaches its peak. Segregation of the biomedical waste is the tactical management policy that enhanced the process cost vividly [26]. Therefore, the training education and awareness problem should be encouraged among the workers in the segregation process. So therefore, the existing operational facilities available in the Hospital waste management (HWM) and municipal solid waste should be strength to reduce the potential impact occurring in the waste management process. In those cases, manufacture of personal protective equipment (PPE) has dominate the waste generation universally. Special waste collection containers should be available nearby areas and in public places to collect the disposal PPEs. Many researches have conducted the disposal and treatment of the facemask recently days. Lot of scope in the recent technologies like microwave treatment, alkaline hydrolysis, steam sterilization, biological treatment [20]. The study conducted in Asia and Africa shows that power generation potential of COVID-19 in pandemic by single use facemask by using incineration technology [27]. Therefore, this technological treatment should become end product that help to build the environment.

5. Conclusion

The study figured out the number of daily single-use facemask has increased in number with respective to the confirmed COVID cases. Additionally, it has been shown some treatments involved in facemask in recent days. Proper and precise method of segregation gives the better outcome. So facemask utilize by the population and healthcare workers is a vital part to manage this pandemic situation. To fight for the novel infectious disease, the government should monitor in all aspects like prevention, guidelines, protocol with some strict implementation that is adversely affect the environmental issues.

References

1. https://www.who.int/publications/i/item/WHO-2019-nCoVIPC-WASH-2020.4 (2020)
2. https://edition.cnn.com/interactive/2020/health/coronavirus-maps-and-cases/
3. https://www.worldometers.info/coronavirus/
4. https://www.mohfw.gov.in/
5. https://www.worldometers.info/coronavirus/country/india/
6. https://prsindia.org/covid-19/cases
7. Alfonso,M.B. ,Arias A.H.Menéndez M C., Ronda A.C, Harte A.C., Piccolo M.C., Marcovecchio J.E, Assessing threats, regulations, and strategies to abate plastic pollution in LAC beaches during COVID-19 pandemic, *Ocean and Coastal Management*, 2021.208 (1): p.105613.
8. Long H Nguyen, Risk of COVID-19 among front-line health-care workers and the general community: a prospective cohort study, *The LANCET Public health,* 2020.5(9): p.475-483.
9. Krithiga P., Sudharsana V., Sribalaji R., Snega C, Covid 19 Pandemic: Assessment Of Knowledge And Attitudes In Biomedical Waste Management Among Health Care Professionals in Tamil Nadu, *Asia-pacific Journal of Health Management,* 2020. 16(3): p.1-11.
10. R. Akhbarizadeh, S. Dobaradaran, I. Nabipour, M. Tangestani, D. Abedi, F . Javanfekr, F. Jeddi, A. Zendehboodi Abandoned Covid-19 personal protective equipment along the Bushehr shores, the Persian Gulf: an emerging source of secondary microplastics in coastlines, *Marine Pollution Bulletin*, 2021.168: p.112386.
11. Sullivan GL., Gallardo JD., Watson TM., Sarp S, An investigation into the leaching of micro and nano particles and chemical pollutants from disposable face masks - linked to the COVID-19 pandemic, *Water Research*, 2021,196: p. 117033.
12. M. Boroujeni, M. Saberian, J. Li Environmental impacts of COVID-19 on Victoria, Australia, witnessed two waves of Coronavirus, *Environmental Science Pollution Research,* 2021, 28(11): p.1-10
13. S. Sangkham, Face mask and medical waste disposal during the novel COVID-19 pandemic in Asia, *Case Studies in Chemical and Environmental Engineering*, 2020, :p.100052.

14. Guidelines for handling, treatment and disposal of waste generated during treatment/diagnosis/quarantine of COVID-19 patients. Revision 4, 17-July-2020 Central Pollution Control Board, Ministry of Environment, Forest & Climate Change, New Delhi (2020).
15. M. Ansari, M.H. Ehrampoush, M. Farzadkia, E. Ahmadi Dynamic assessment of economic and environmental performance index and generation, composition, environmental and human health risks of hospital solid waste in developing countries; A state of the art of review, *Environment International*,2019,132: p.105073.
16. A. Majumder, A.K. Gupta, P.S. Ghosal, M. Varma, A review on hospital wastewater treatment: a special emphasis on occurrence and removal of pharmaceutically active compounds, resistant microorganisms, and SARS-CoV-2, *Journal of Environment Chemical Engineering*, 2021, 9 (2): p.104812
17. https://cpcb.nic.in/uploads/Projects/Bio-Medical Waste/COVID19_Waste_Management_status_August2020.pdf.
18. https://cpcb.nic.in/uploads/Projects/Bio-Medical-Waste/COVID19_Waste_Management_status_April-May_2022.pdf.
19. A. Richter, K.T.W. Ng, H.L. Vu, G. Kabir Waste disposal characteristics and data variability in a mid-sized Canadian city during COVID-19, *Waste Management*, 2021: p.49-54.
20. Ojha PC., Satpathy SS., Ojha AK., Sukla LB., Pradhan D, Overcoming challenges due to enhanced biomedical waste generation during COVID-19 pandemic, *Science Total Environment,* 2022: p. 832:155072
21. C. Nzediegwu, S.X. Chang, Improper solid waste management increases potential for COVID-19 spread in developing countries, *Resource Conservation Recycling*, 2022, 161: p. 104947.
22. Sarawut Sangkham., (2020). Face mask and medical waste disposal during the novel COVID-19 pandemic in Asia. *Case Studies in Chemical and Environmental Engineering,* 2020, 2: p.100052
23. A. Tripathi, V.K. Tyagi, V. Vivekanand, P. Bose, S. Suthar Challenges, opportunities and progress in solid waste management during COVID-19 pandemic, *Case Studies in Chemical and Environmental Engineering,* 2020:p.100060.
24. B.N. Kulkarni, V. Anantharama, Repercussions of COVID-19 pandemic on municipal solid waste management: Challenges and opportunities, *Science Total Environment*, 2020, 743:p. 743140693.
25. F.G. Torres, E.G. De-la-Torre Face mask waste generation and management during the COVID-19 pandemic: an overview and the Peruvian case, *Science Total Environment*, 2021, 786: p. 147628.
26. D'Souza B.C., Seetharam A.M., Chandrasekaran V., Kamath R. Comparative analysis of cost of biomedical waste management across varying bed strengths in rural India, *International Journal of Health Management*, 2018,11:p. 38–43.

27. Cudjoe D., Wang H., Zhu B, Thermochemical treatment of daily COVID-19 single-use facemask waste: Power generation potential and environmental impact analysis, *Energy,* 2022, 249:p. 123707.

Investigation of the Optimum Shotcrete Mixing Ratio for Slope Protection using Fly Ash

[1]*Thang* Kel Win [2]*Nur Irfah* Mohd Pauzi,

[1] Institute of Energy Infrastructure, College of Engineering, University Tenaga Nasional Berhad (UNITEN), Putrajaya Campus, Jalan Kajang, Puchong, 43000, Kajang, Selangor, Malaysia.
[2] Civil Engineering Department, College of Engineering, University Tenaga Nasional Berhad (UNITEN), Putrajaya Campus, Jalan Kajang, Puchong, 43000, Kajang, Selangor, Malaysia

Abstract Shotcrete has now grown in popularity and can replace cast-in-place (CIP) concrete in applications such as retaining walls and slope stabilisation. However, due to the problems like cracking due to shrinkage and the durability concern, this practise is still limited. So, cementitious material such as fly ash is used in this situation to increase the strength of the concrete and also to make sure it is high durability and safe to use in construction. Fly ash can be known as a coal combustion residual which is commonly occur in thermal power plant and it had been widely used as an alternative material during the production of Portland cement concrete. Supplementary cementitious materials is widely used in the mixing of concrete due to its unique properties which can harden the concrete via its hydraulic and pozzolanic activity. A series of experiments will be conducted to make sure that the fly ash that had been mixed with Portland cement is safe to use and also can increase the compressive strength of the cement. Throughout the experiments, a series of data will be display with the addition of 5%,10%,15%,20%,25%,30%,35% of pozzolans in the cement mixing ratio. The addition of fly ash during the mixing of concrete will indirectly reduce the cost of construction and also the rate of pollution to the environment as it is being classified as a scheduled waste in Malaysia according to the Environmental Quality Act 1974.This research will mainly focus on the performance of shotcrete with the addition of cementitious material which is fly ash as an alternative product compare to the tradition concrete and compare the differences in the achievement of shotcrete with and without the addition of fly ash so that the comparison is more clearly and more understandable. Next, the utilization of the nominal aggregates will be 10-20mm.The amount of fly ash added to the mixing of concrete will be in the range of 5% to 35%.

1 Introduction

After so many studies that had been done on the effect of fly ash in concrete, thus there is still have limitations or problems that need to be solved to make sure that the concrete is safe to be used. Next, the results of adding pozzolans such as fly ash on the mechanical properties of shotcrete is still in a unknow space and researchers should pay more attention to this area. (Ai J, 2007)In order to improve the reputation of our nation, it is very vital to make sure that all the scheduled waste is being used well instead of throw it aside. So, the unknown space mentioned above is actually a very good research gap for researchers to explore more knowledge in this field. Indirectly, with the usage of fly ash in shotcrete will save up a lot of construction cost as it is being classified as a scheduled waste in Malaysia. It will be a very big improvement to our nation with the success of utilizing scheduled waste in shotcrete.

Next, fly ash is being produced through the combustion of coal in the power plant and it carries around 30-45% of ash content. All of fly ash are required to be disposed through landfill or ponds and if construction field can utilise these resources properly, it will indirectly save up a lot of cost too. On the other hand, with the poor management of these scheduled waste, it may cause pollution to water and environment and indirectly it will harm the health of population. It contains a lot of toxic components that may bring harm to the environment. The ponds where fly ash is disposed should be well managed to prevent it from becoming the main sources of air and water pollution. Central Electricity

© The Authors, published by EDP Sciences. This is an open access article distributed under the terms of the Creative Commons Attribution License 4.0 (https://creativecommons.org/licenses/by/4.0/).

Authority stated that the number of fly ash generation has increased a lot which is initially from 70 million ton to 170 million ton.

2 Research Objective

This research will mainly focus on the performance of shotcrete with the addition of cementitious material which is fly ash as an alternative product compare to the tradition concrete. The research will also compare the differences in the achievement of shotcrete with and without the addition of fly ash so that the comparison is more clearly and more understandable. Next, the utilisation of the nominal aggregates will be 10-20mm.The percentage of fly ash added to the mixing of concrete will be 5%,10%,15%,20%,25% and 30%. So, the following research objectives will be achieved throughout the laboratory experiments:

1. Investigate the fresh properties of the concrete (slump value)
2. Evaluate the mechanical properties of the shotcrete with and without the addition of fly ash.
3. Determine the optimum mixing ratio of shotcrete with the addition of fly ash by using statistical analysis software.

3 Summary of Literature Review

Table 1. Summary of Literature Review

No	Author	Title	Findings
1	Zhou Ming Xue, Ma Zhemei, Xiao Xing heng (Zhou Mingxue, 2009)	Study on the effect of fineness and dosage of fly ash on mechanical properties of concrete	-The smaller the particle size of fly ash is, the more significant its activity effect is, so the greater the optimal mixing amount of fly ash in concrete is. -When the particle size of fly ash is less than 5μm, the optimal dosage of fly ash in concrete is 30% in order to improve the 60d compressive strength.
2	Bin Wu, Mingming Chai, Shijia Li and Jianping Yang (Bin Wu, 2022)	Effect of Fly Ash Content in Shotcrete on Mechanical Properties	-Combined with the change of concrete strength, to guarantee the structural safety, the proportion of fly ash is no more than 15%. Otherwise, the strength loss will be too serious for a safe construction structure.
3	Wei Lu, Jiawei Wang, Zhiqiang Peng, Biaozhi Zhang (Wei Lu, 2018)	Research on influences of fly ash content on mechanical Properties of shotcrete	-Under the 28 days standard curing condition, the compressive strength, tensile strength and flexural strength of shotcrete are negatively correlated with fly ash content, that is, the higher the fly ash content is, the greater the loss of the mechanical strength is. -For compressive strength, the adding content of fly ash instead of cement should not exceed 15% in actual engineering construction.
4	Michael Thomas (Thomas, 2003)	Optimising the use of fly ash in concrete	-The optimum fly ash content will vary on a case-by case basis. Fly ash contents of up to 50% may be suitable for most

			elements provided the early-age strength requirements of the project can be met and provided that adequate moist-curing can be ensured. -For flatwork, the level may be dictated by finishing requirements.
5	Zhen Hua Duan, Chi Sun Poon (Zhen Hua Duan, 2014)	Effect of Fly Ash on the Mechanical Properties and Microstructure of Cement Stabilized Materials with 100% Recycled Mixed Aggregates	-The incorporation of fly ash can significantly improve the drying shrinkage performance, increase the mass loss, and reduce abrasion resistance of CSM with 100% RMA. -When the replacement rate of fly ash was 10%, 20%, 30%, and 40%, the total drying shrinkage coefficient was reduced by 9.3%, 19.4%, 34.3%, and 63.21%, respectively, compare to baseline (0%).
6	Mohammed Seddik, Bencheikh (Mohammed Seddik, 2009)	Impact on Mechanical Properties of SFRC for Cement Replacement Using Fly Ash	As the Fly ash content is increase gradually in SFRC, the compressive strength decreased gradually. The compressive strength shows a decreasing trend with increase in fly ash percentage.
7	Yuji Samaru and Koji Sugiyama (Yuji Samaru, 2005)	The effect of bottom ash replacement as fine aggregate on the property of shotcrete	- The mixture of shotcrete used bottom ash as a fine aggregate can be reduced the unit weight by approximately 20% of normal shotcrete. - The slump flow of shotcrete used bottom ash as a fine aggregate was less than shotcrete used sand as a fine aggregate. Shotcrete mixed with a middle size of aggregate gave the highest slump flow value. - Sagging of shotcrete used bottom ash as a fine aggregate was less than the normal shotcrete at the same thickness and inclination angle.
8	Arnie Rosenberg (Rosenberg, 2010)	Using Fly Ash in Concrete	- Since fly ash particles are spherical and in the same size range as Portland cement, a reduction in the amount of water needed for mixing and placing concrete can be obtained. In precast concrete, this can be translated into better workability, resulting in sharp and distinctive corners and edges with a better surface appearance. - Fly ash is considered to have met C618's requirements if the 7- or 28-day strength of a sample with 20 percent fly ash reaches 75 percent of the control strength in an ASTM C109 test.
9	Prajwala S. Chachadi1, Prasanna Patil	A Review on Fly Ash Slope Stability	- The objective of the study is to check the feasibility of using plastic recycled polymers with fly ash in construction of steep slopes so that a

	(Prajwala S. Chachadi, 2021)		innovative methodology could be adopted for its utilization. The experimental analysis and validation using finite element method for the methodology.
10	J.N.Jha, A.K.Choudhary, K.S.Gill ,S.K.Shukla (J.N.Jha, 2013)	Bearing Capacity of a Strip Footing Resting on Reinforced Fly Ash Slope: An Analytical Approach	- There is difficulty due to poor loadbearing capacity of fly ash, when footings rest on the top of the fly ash fill slope; but inclusion of polymeric reinforcements as horizontal sheets within the fill may be advantageous in improving the load-bearing capacity of reinforced fly ash slope

4 Research gap

Ordinary Portland cement concrete (OPCC) is one of the frequently adopted construction materials worldwide. With the increase in Portland Cement production, environmental pollution will become increasingly serious problems. Using fly ash in concrete reduces cracking, permeability, and bleeding, creating a dense, high-durability concrete that is resistant to sulphates and alkali-aggregate reactions.

However, using fly ash in concrete does come with potential problems. Substituting fly ash for Portland cement in a concrete mixture leads to longer drying and curing times. When used in cold climates, mixtures with high levels of fly ash can also encounter issues with slow strength development. Therefore, the setting time of the fly ash concrete need to be further studies. Next, the disposal of fly ash concrete also needs to be planned well as fly ash concrete is not environmentally friendly. It is a waste by product of combusted coal, and it's generally considered an environmental hazard. In construction field, different percentage of fly ash is mixed in concrete and the percentage of the fly ash should be standardise in certain elements. For example, the percentage of fly ash in the concrete for slabs should not more than 15% because of the safety factor.

5 Methodology

5.1 General

The methodology of the research is shown in Figure 1. The methodology starts from introduction of the main components, then literature review to identify the most ideal percentage of the fly ash to be included during the concrete mixing process. Aggregates were collected at Kajang Rocks and fly ash were collected at Manjung, Perak.

5.2 Specimen preparation

The samples were prepared according to the addition amount of fly ash and also the curing time . The cement with 0% of fly ash acts as control variable and to compare the results with additives of fly ash. The whole process of preparing the samples consists of five basic materials which are aggregates, cement, fly ash , water and superplasticizer.

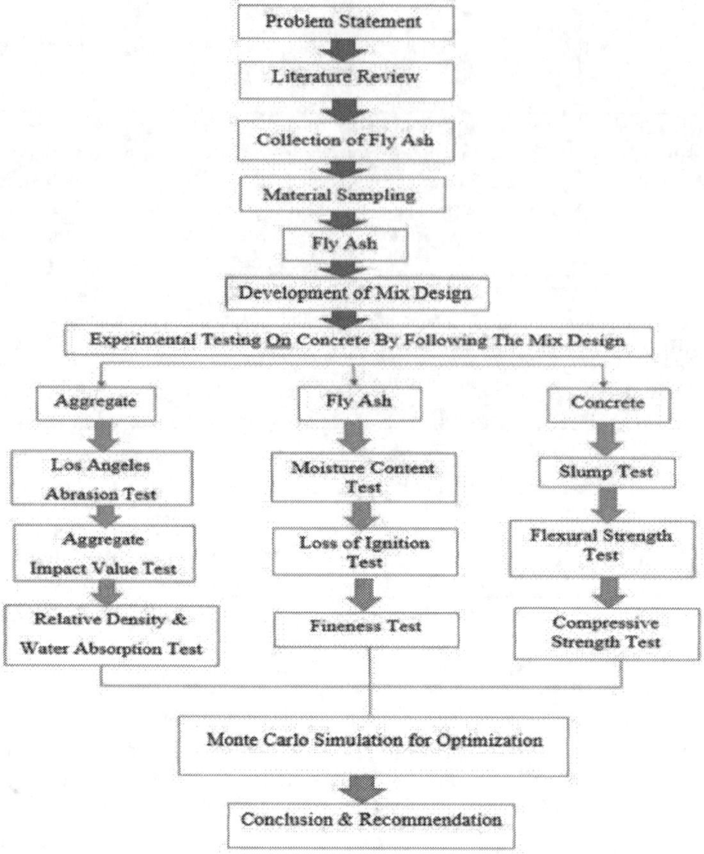

Fig.1. Flowchart Of The Research Methodology

6 Preliminary Results

6.1 Experiment for aggregate

6.1.1 Sieve Analysis (ASTM C136)

Table 2. Results of sieve analysis

Grading	Mass of sample before test (g) (m_1)	Mass of oven-dried sample retained on sieve size 1.70mm (g) (m_2)	Mass of sample passing sieve size 1.70mm (g)	LAA value(%)
A	5010	3628.5	1381.5	27.57
B	5000	3465.0	1535.0	30.70
C	5000	3336.5	1663.5	33.27

6.1.2 Aggregate Impact Value(BS812:Part 112:1990)

Table 3. Results of Aggregate Impact Value

Sample	1	2	3
Weight of cylinder steel cup (g)	3034.6	3034.6	3034.6
Weight of cylinder steel cup + sample (g)	3679.1	3675.9	3680.6
Weight of sample (g) [M1]	644.5	641.3	646
Weight of sample passing 2.36mm sieve (g) [M2]	96.7	98.4	93.3
Weight of sample retained on 2.36mm sieve (g) [M3]	543	540.8	534.2
Aggregate impact value (%)	14.973	15.344	14.443
Average aggregate impact value (%)	14.92		

6.1.3 Relative Density and Water Absorption of Aggregates [ASTM C128]

Table 4. Result of Relative Density and Water Absorption of Aggregates

Description		Sample		
		1	2	3
Weight of sample, g		1000.5	1000.0	1001.0
Weight of saturated & dry sample, g [A]		984.5	1004.5	970.5
Weight of basket + sample + water, g [B]		1380	1380	1370
Weight of basket + water, g [C]		770	760	770
Weight of oven dry sample, g [D]		960.5	972.0	955.0
Weight of basket, g		900	900	900
Specific gravity		2.6	2.5	2.6
Apparent specific gravity		2.7	2.8	2.7
Water absorption, Percentage Dry Weight, %		2.5	3.3	1.6
Average Value	Specific gravity	2.6		
	Apparent specific gravity	2.7		
	Water absorption, %	2.8		

6.2 Experiment for Fly Ash

6.2.1 Moisture content test of fly ash (ASTM C311-11)

Table 5. Result of moisture content test of fly ash

Sample	Weight of dish (g)	Initial mass of sample (g) [C]	Initial mass of dish and sample (g) [A]	Mass of oven dried sample + dish (g) [B]	Final mass of oven dried sample (g)	Mass loss of sample during drying (g)	Free-moisture content of sample (%)
A	2.5	100	102.5	100.00	97.5	2.5	2.5
B	2.5	100	102.5	102.50	100.0	0.0	0.0
C	2.5	100	102.5	102.50	100.0	0.0	0.0

6.2.2 Loss on Ignition of Fly Ash (ASTM C311-13)

Table 6. Result of loss of ignition of fly ash

Sample	Weight of crucible (g)	Initial weight of sample(g) [C]	Weight of crucible + sample(g) [A]	Weight of crucible +ignition sample(g) [B]	Final weight of ignition sample(g)	Loss on ignition of (%)
A	61.150	100.221	161.371	157.864	96.714	3.50
B	53.955	100.200	154.155	150.876	96.921	3.27
C	61.840	100.550	162.390	158.751	96.911	3.62
D	61.871	100.659	162.530	159.099	97.288	3.41

6.2.3 Fineness test of fly ash (ASTM C311-19)

Table 7. Result of finesness test of fly ash

Sample	1	2	3
Weight of cylinder steel cup (g)	3035	3035	3035
Weight of cylinder steel cup + sample (g)	3757.0	3743.0	3729.5
Weight of sample (g) [M1]	722.0	708.0	694.5
Weight of sample passing 2.36mm sieve (g) [M2]	47.0	39.0	41.5
Weight of sample retained on 2.36mm sieve (g) [M3]	672.0	655.5	648.5
Aggregate impact value (%)	6.51	5.51	5.98
Average aggregate impact value (%)	6.00		

https://doi.org/10.1051/matecconf/202338402004

6.3 Experiment for Concrete

6.3.1 Slump test (ASTM C143)

Table 8. Result of slump test

Type of Concrete	Slump Value(mm)
Normal Concrete M30	105
M30 concrete + 5% Fly Ash	105
M30 concrete + 10 % Fly Ash	110
M30 concrete + 15 % Fly Ash	111
M30 concrete + 20 % Fly Ash	115
M30 concrete + 25 % Fly Ash	120
M30 concrete + 30 % Fly Ash	110

6.3.2 Flexural Strength of Concrete (ASTM C78)

Table 9. Result of flexural strength of concrete

Type of Concrete	Flexural Srength at 7 days (Mpa)	Flexural Srength at 14 days (Mpa)	Flexural Srength at 28 days (Mpa)
Normal Concrete M30	3.85	3.64	4.50
M30 concrete + 5% Fly Ash	4.43	3.95	4.09
M30 concrete + 10 % Fly Ash	3.89	4.8	
M30 concrete + 15 % Fly Ash	2.15	3.41	4.38
M30 concrete + 20 % Fly Ash	3.12	2.81	4.82
M30 concrete + 25 % Fly Ash	1.89	3.03	4.74
M30 concrete + 30 % Fly Ash	2.36	1.88	4.06

6.3.3 Compressive Strength Test (ASTM C39)

Compressive strength is the ability of material or structure to carry the loads on its surface without any crack or deflection. A material under compression tends to reduce the size, while in tension, size elongates.

Table 10. Result of compressive strength test

Type of concrete	Curing Day	Compressive Strength (MPa)
Normal Concrete M30	3 days	23.63
	7 days	24.06
	28 days	33.03
M30 concrete + 5 % Fly Ash	3 days	22.1
	7 days	28.67
	28 days	32.27
M30 concrete + 10 % Fly Ash	3 days	19.32
	7 days	22.7
	28 days	22.43
M30 concrete + 15 % Fly Ash	3 days	15.47
	7 days	22.76
	28 days	29.94
M30 concrete + 20 % Fly Ash	3 days	14.27
	7 days	21.36
	28 days	28.05
M30 concrete + 25 % Fly Ash	3 days	18.47
	7 days	23.41
	28 days	20.36
M30 concrete + 30 % Fly Ash	3 days	10.17
	7 days	7.88
	28 days	17.86

6.4 Monte Carlo Simulation

Monte Carlo simulation is a statistical technique by which a quantity is calculated repeatedly, using randomly selected "what-if" scenarios for each calculation. Though the simulation process is internally complex, commercial computer software performs the calculations as a single operation, presenting results in simple graphs and tables. These results approximate the full range of possible outcomes, and the likelihood of each. When Monte Carlo simulation is applied to risk assessment, risk appears as a frequency distribution graph similar to the familiar bell-shaped curve, which non-statisticians can understand intuitively. The technique is used by professionals in such widely disparate fields as finance, project management, energy, manufacturing, engineering, research and development, insurance, oil & gas, transportation.

With the application of Monte Carlo Simulation, the optimum mixing ratio of fly ash in concrete can be determined. With the results, we can identify what is the maximum content of fly ash in concrete and thus it saves a lot of time during construction and indirectly reduce the environmental impact as all the fly ash was mixed with concrete for construction uses.

7 Conclusions

In conclusion, fly ash plays a vital roles in the mixing of concrete. A series of test had been conducted to test the physical and chemical properties of all the raw materials and finally the compressive strength of the concrete. It can be said that it is a successful research as the fly ash concrete can achieve the ideal compressive strength. With the addition of fly ash in the mixing of concrete, it can also indirectly reduce the pollution to environment as fly ash is scheduled as waste schedule in Malaysia according to rules and regulations. Next, with the utilisation of this technology will increase the reputation of our nation too.

The smaller the particle size of fly ash is, the more significant its activity effect is, so the greater the optimal mixing amount of fly ash in concrete is. The optimum mixing ratio based on Monte Carlo Simulation and test shows that the concrete with addition of 15%-20% fly ash has the highest compressive strength.

8 Acknowledgment

The author would like to acknowledge the University Tenaga Nasional for supporting the research under the TNB Seed Fund BOLD grant J510050002/2022028 and also to the Civil Engineering Department, UNITEN for providing the facilities for library resources.

9 References

1. Ai J, D. S. (2007). The performance of RCC with large volume high calcium fly-ash. 33-35.
2. Apha Sathonsaowaphak, P. C. (2009). Workability and strength of lignite bottom ash geopolymer mortar. *Journal of Hazardous Materials*, 44-50.
3. Bin Wu, M. C. (2022). Effect of Fly Ash Content in Shotcrete on Mechanical. *Journal of Physics: Conference Series*, 3-7.
4. Cao J Liu, L. Z. (2017). Study on developing mechanical properties. *Journal of Nanchang Institute of Technolog y 36*, 62-66.
5. Cui Z, Z. X. (2019). Effects of different curing environments on strength and. 65-76.
6. Eugeniusz Koda, P. O. (2011). Slope erosion control with the use of fl y-ash and sewage sludge. 101-111.
7. Hou W, Y. D. (2017). Study on performance of concrete with different content of fly ash. *Eady mixed Concrete (100*, 100-102.
8. J.N.Jha, A. K. (2013). Bearing Capacity of a Strip Footing Restingon Reinforced Fly Ash Slope. 354-366.

9. Li H, Y. C. (2017). Effect of fly ash on mechanical properties of concrete. *Guangdong Building Materials 33*, 15-17.

10. Mohammed Seddik, B. (2009). Properties of concrete reinforced with different kinds of industrial waste fibre materials. 3196-3205.

11. Prajwala S. Chachadi, P. P. (2021). A Review on Fly Ash Slope Stability. 1-5.

12. Rosenberg, A. (2010). Using Fly Ash in Concrete. 1400-1410.

13. Thomas, M. (2003). Optimizing the Use of Fly Ash in Concrete. 29-33.

14. Tushar Vasant Salunkhe, S. M. (2015). Stabilisation of Fly Ash Slope using Plastic Recycled Polymer. 2-5.

15. Wei Lu, J. W. (2018). Research on influences of fly ash content on mechanical properties of shotcrete. 2-22.

16. Yang X, P. Y. (2004). periment and Study on Ash Sprayed-concrete in Laboratory [J. 8-9.

MATEC Web of Conferences **384**, 02005 (2023) https://doi.org/10.1051/matecconf/202338402005
SPICE 2023

Investigation on Performance of Heritage Masonry Structures

Sreenivas Muthukumaraswamy Kamalakannan[1], *Sravan Ashwin* Anandamurali[2], and *Arunachalam* Parthasarathy[3]
[1, 2, 3.] *Student, College of Engineering Guindy, Anna University India – 603110*
E-mail: srinivas.kamalakannan@gmail.com
E-mail: arunpartha2019@gmail.com
E-mail: ashwinsa25@gmail.com

Abstract

A structural assessment study on the "Humayun Palace" in Tamil Nadu, a cultural heritage structure, constructed in the early 1700's was done. Based on the original design documentation collected through records, a static analysis model of the building was generated. This structure is a 3-storey building and details of all the provisions from the slab system to the roof tiles were made. The results developed by these models, aimed at fully understanding the original design concept of the various members, as well as at evaluating their current static and seismic safety conditions, are reported in this paper. This analysis was examined using E-TABS software, whose geometric limitations lead to minor encumbrances. Some complex geometric façade features were approximated to convenience and overcome. The results of the analyses highlight safe conditions and good performance objectives in general, but for some important exceptions. The overall satisfactory performance may be attributed to the overdesigning with excessive safety factors carried out by builders of that century, enabling the structure to withstand conditions way past its intended life. However, the question as to how well the structure will hold good in the future is hypothetical. It is for the same reason that the Public Works Department has allocated nearly 96 crore rupees for the rehabilitation of the same. This investigation is synchronous with the objectives of the said rehabilitation project and has established an exhaustive analysis of the structural health of the structure.

Keywords: Un-reinforced masonry, Rehabilitation, Response reduction factor, Importance factor, Retrofitting

1. Introduction

1.1 General

World-wide experience from past earthquakes reveals the particular

© The Authors, published by EDP Sciences. This is an open access article distributed under the terms of the Creative Commons Attribution License 4.0 (https://creativecommons.org/licenses/by/4.0/).

vulnerability of un-reinforced masonry structures to seismic action. Hence, these structures represent a severe threat to human and other resources in the case of similar future seismic events. This requires an urgent assessment of existing buildings in terms ofthe strength, expected performance and safety of existing buildings during earthquake as well as for carrying out the necessary rehabilitation. The Chepauk Palace comprises two blocks—the northern block is known as Kalas Mahal while the southern block is known as Humayun Mahal. Located within a 117 acre campus, the Humayun Mahal is spread over 66,000 square feet and has ventilation on the terrace and a connecting corridor to the Kalas Mahal. Fig 1. exhibits the structure's original physical appearance during time of construction. This investigation was focused on Humayun Mahal specifically.

Fig 1. Original Construction of the Chepauk Palace.

Fig 2. Renovation of the Structure during fire accident in 2016.

1.2 Need for Study

Humayun Mahal has suffered decades of neglect: languishing in dilapidated condition for years and later ravaged by fire. Due to such improper maintenance and the recent fire accident deteriorating the structure even further, the current strength, stability and durability of the structure is questionable. For the same reason the Public Works Department of India has decided to allocate Rupees 95.98 crores for the renovation of the same. Figure 2 shows the renovation that took place for the structure after the devastating fire accident. The significance of architectural heritage and the desirability of conserving it are now more readily recognized, asnoticed in the largely increasing public concern in favor of preserving and enhancing the familiar local scene.

1.3 Objective

The main agenda in this project is to model and analyze the said structure. The intention for experimentation is to assess the performance of the structure. The structure was built 252 years ago and hence, there is a pertinent need to model the structure using the E-tabs software and adopt the static analysis model for the same. Relevant loads such as dead load, live load, wind load and seismic load are applied on the model in accordance to currently prevalent Indian Standard Codes. In doing so, an assessment on to what extent the structure will fare in today's context can be obtained. Through this investigation, an exhaustive report on the health of the structure and possible retrofitting or rehabilitation for the sections of the structure which do not show promising results is done.

2. Literature Study

2.1 General

Ancient monumental structures are complex structures but have not been engineered to undergo the many transformations they encounter during their lifeand often present lack of connections among the structural elements. For this purpose, extensive research and investigation on this agenda has garnered popularity over the years. The following research papers discussed in the below section are the basis and inspiration of this study.

2.2 Literature Review

Ahmad et al (2014) [1] have investigated the structural inadequacy of a century old unreinforced masonry heritage building, to withstand seismic loads and have extended that such similar structures require an urgent assessment in terms of the strength, expected performance and safety. In this study they have investigated and experimented to evaluate the seismic performance of a 137-year-old unreinforced masonry heritage school building located at Aligarh Muslim University, Aligarh (seismic zone IV). This given structure is a load bearing structure. The seismic analysis of the building is done using equivalent static method as described in the code IS 1893(Part I): 2002. Natural frequencies and mode shapes of the building have been obtained through modal analysis approach using STAAD Pro. The building is modelled using finite element technique and its seismic evaluation is carried out assuming a homogeneous and nonlinear behavior of the material. The building is subjected to different peak ground acceleration (PGA) levels (0.1g, 0.2g, 0.3g, 0.4g) as input ground motion. The portions around the openings were found to be highly vulnerable in all cases.

Miceikaitė and Parasonis [2] have investigated the representative palace of Biržai castle. The palace was rebuilt and tailored to become a museum and library. All these rebuilding works were done without sufficient geotechnical data. The purpose is to find out the explanation for the exterior walls splitting and present solutions to stabilize the current situation. The model was simulated using STAAD Pro and Finite Element Analysis was performed for the same. Two methods adopted were (a). Modelling the wall without taking splits into account in the current wall, (b). Modelling the wall with its natural splits. It is inferred that splits are influencing both stresses values and their distribution in the wall. Authors conclude that an improper remains analysis of masonry structures was done, leading to the difficulty in ascertaining the distinction between original and new wall places. For that reason, it is hard to decide on masonry loadbearing values. Computing analysis results assert that a hypothesis for compressive strengths of 1.6-2.6 MPa leads to over-limit principal stresses without taking the splits into account. The results highlight that wall stresses and deflection results are influenced by thrust loads and splits places. This information is useful to decide on suitable walls consolidation techniques.

Saloustros et al (2020) [3] have presented the systematic use of numerical analysis as a tool for addressing some of the most common challenges encountered in the structural analysis of complex historical masonry structures. Conservation of built cultural heritage is a challenging task that relies on the implementation of a multidisciplinary scientific approach based on inspection, structural analysis, and monitoring. This study deals with the application of FE macro-modeling analysis for the structural assessment of complex historical masonry structures against long-term deformation, past historical alterations, and seismic actions. The first part is devoted to the presentation of a FEM model based on continuum mechanics theory. This model is able to reflect mechanical phenomena typically encountered in masonry structures. The second part presents two applications of the FEM on two important monuments in Spain: the cathedral of Mallorca and the church of the Poblet monastery, a UNESCO World Heritage Site. The third part includes a probabilistic analysis for assessing the effect of uncertainties related with material and numerical properties on the seismic response of the church of the Poblet monastery. The proposed mechanical model for masonry is able to simulate the time-dependent strain accumulation in the material because of long-term exposure to constant stress, including the description of tensile and compressive damage. It also elucidated on the use of numerical tools for structural analysis and their combination with historical investigation, in-situ inspection, and monitoring activities for two important large masonry historical structures in Spain. A numerical strategy based on the use of a macro-modeling approach was chosen as a way to reduce the computational cost and to permit the efficient analysis of the large-scale structures. The constitutive model plays an important role in this strategy; it was chosen to properly represent the distinct nonlinear tensile and compressive behavior of masonry, as well as its time-dependent response. The application of this methodology to case studies permitted the interpretation of their current pathology and allows to take optimum decisions of future studies, such as the definition of monitoring activities.

Dizhur et.al. [4] have observed the earthquake damage to masonry buildings and churches as a result of the 22nd of February 2011 Christchurch earthquake. The study focused on investigating commonly encountered failure patterns and collapse mechanisms. A detailed study was done based on the observations that were made on the performance and the deficiencies that contributed to the damage formed in unreinforced clay brick masonry (URM) buildings, unreinforced stone masonry buildings, reinforced concrete masonry (RCM) buildings. This was achieved by investigating the failure patternsand collapse mechanisms that were commonly encountered. The shapes and modelling of the structure is done by suitable software. The modelling of the building was carriedout by finite element analysis. From the above observation, details are provided of retrofit techniques that were implemented within relevant Christchurch URMbuildings prior to the 22nd of February earthquake. A very brief outline on the possible retrofit strategies and repair techniques for stone masonry buildings was also presented.

2.3 Summary of Literature

From these references, it was construed as to how various investigations on various heritage structures were conducted. The usefulness of the results of a model simulation in aiding the structural health of a structure was assessed. The various retrofitting and rehabilitation strategies given help us understand that such results from simulations can be implemented in real time work, which is extremely efficient and accurate. To dwell more on the same lines,a similar investigation is undertaken on the Humayun Palace.

3. Methodology

Work methodology is given in a flowchart as follows:

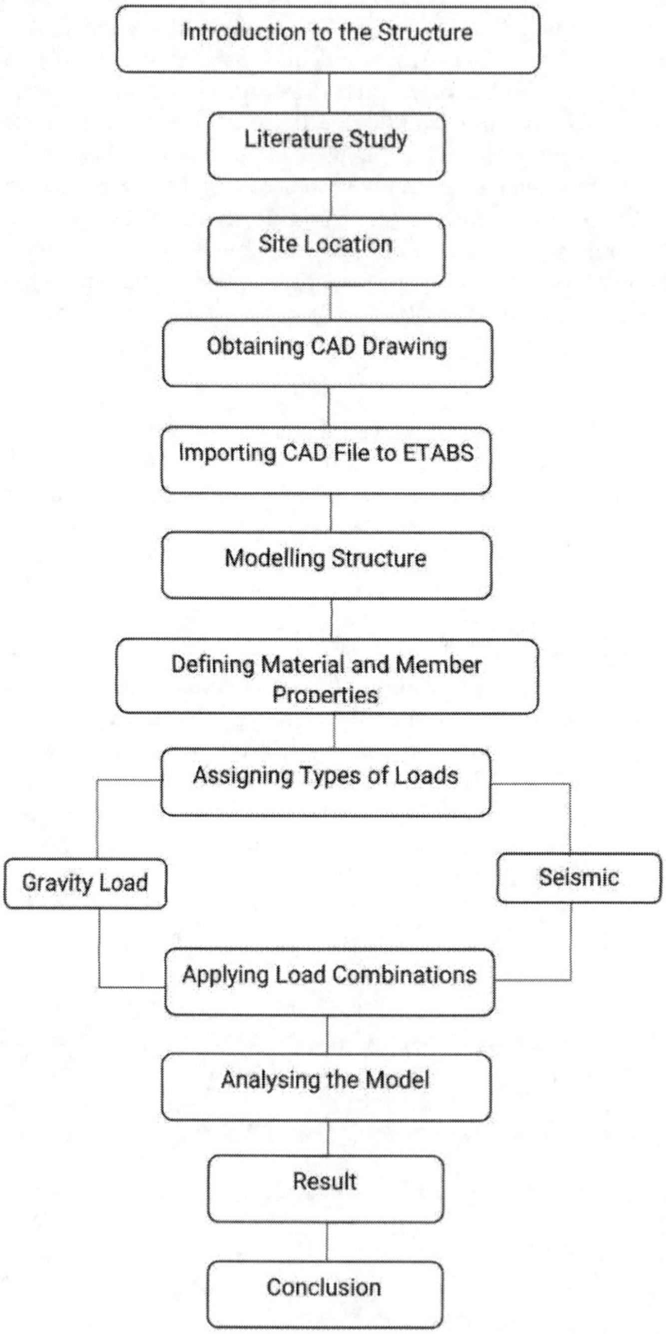

3.1. Layout of the Structure

The plan layout (various views and sections)of the Humayun palace is shown in Fig.3 a-f.

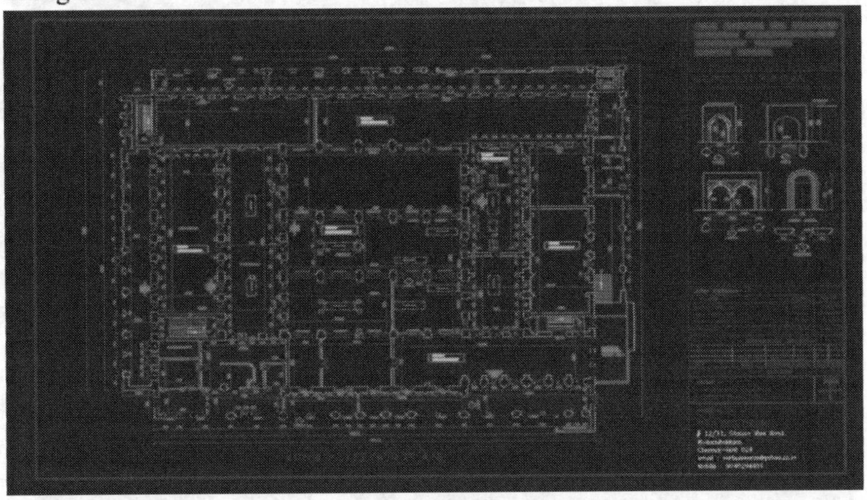

Fig.3.a Shows the Plan layout of the Humayun Palace in the format of a CAD drawing.

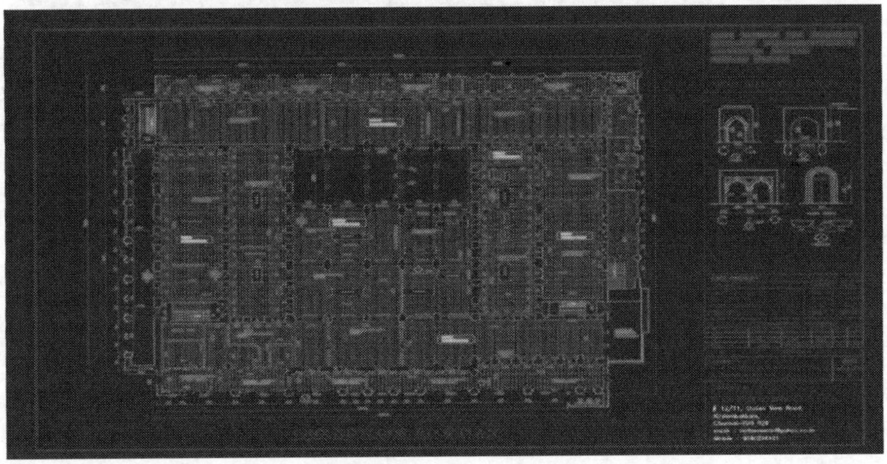

Fig.3.b. Shows the Plan layout of the Humayun Palace in the format of a CAD drawing.

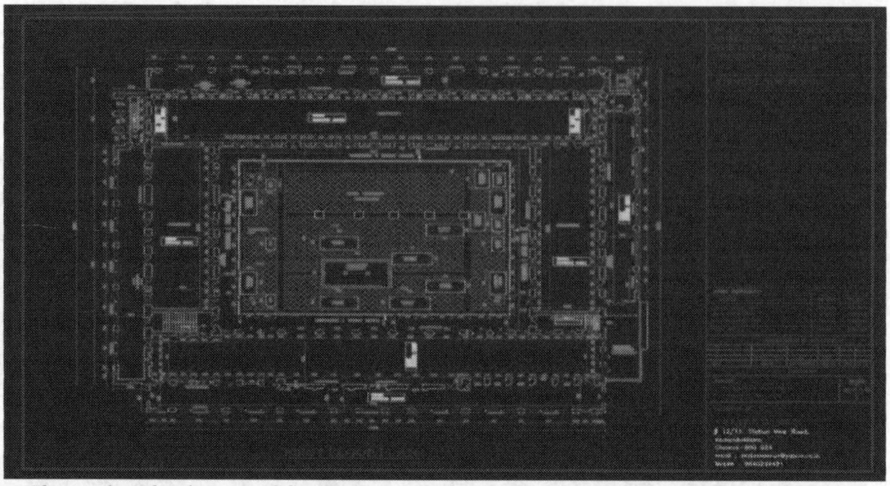

Fig.3.c. Shows the Plan layout of the Humayun Palace in the format of a CAD drawing.

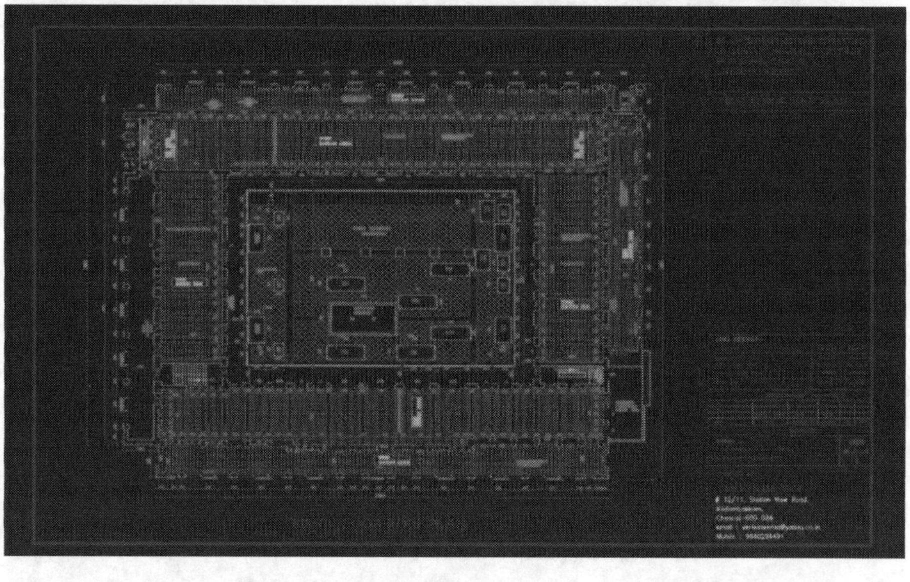

Fig.3.d. Shows the Plan layout of the Humayun Palace in the format of a CAD drawing.

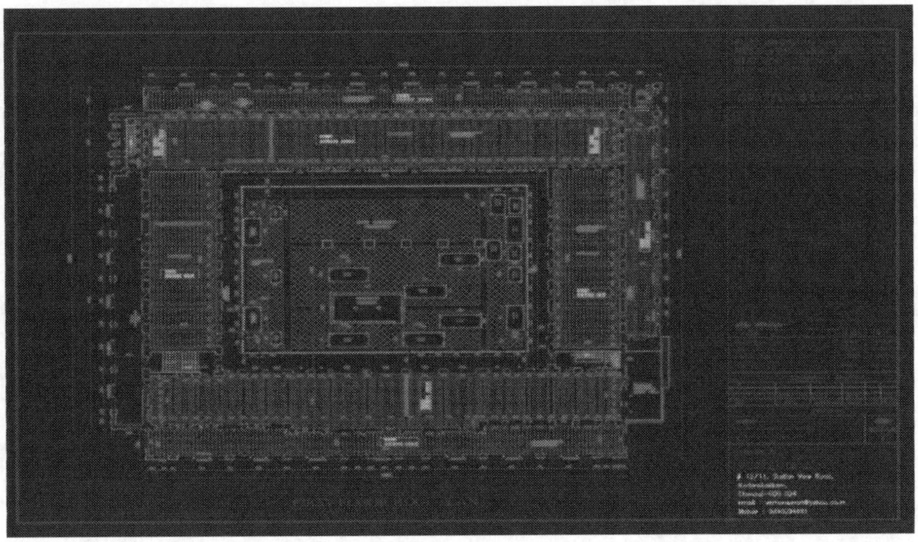

Fig.3.e. Shows the Plan layout of the Humayun Palace in the format of a CAD drawing.

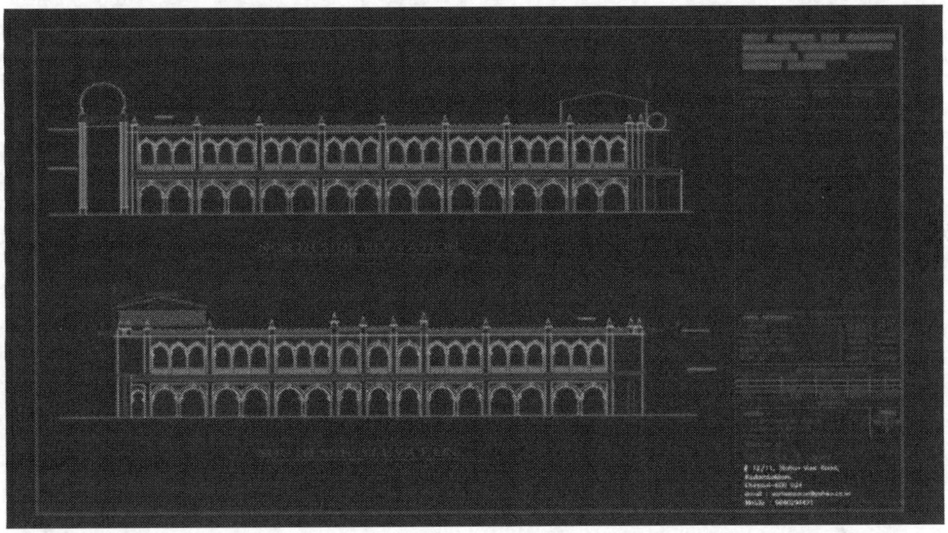

Fig.3.f. Shows the Plan layout of the Humayun Palace in the format of a CAD drawing.

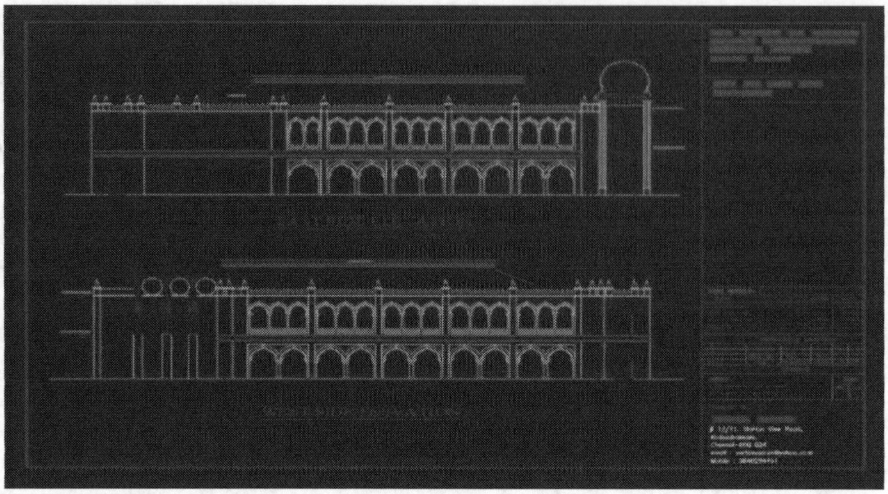

Fig.3.g. Shows the Plan layout of the Humayun Palace in the format of a CAD drawing.

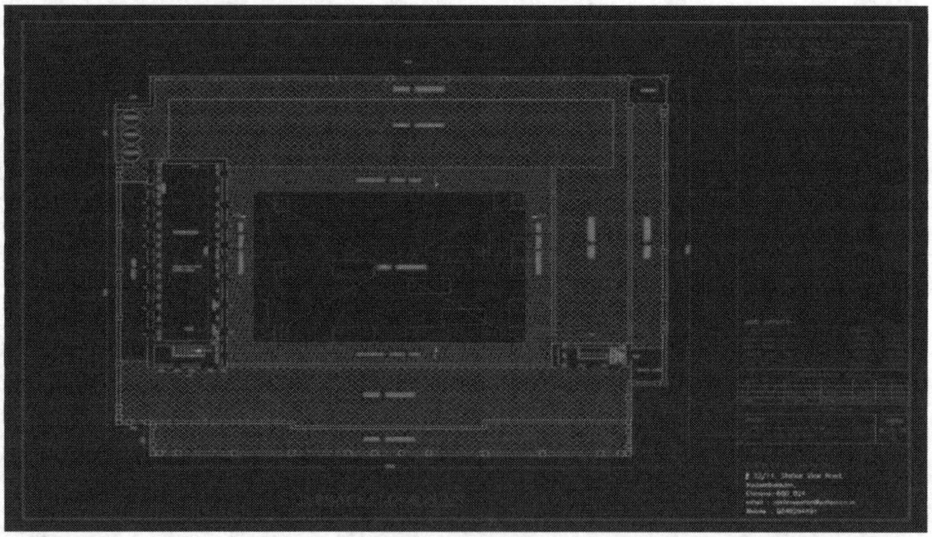

Fig.3.f. Shows the Plan layout of the Humayun Palace in the format of a CAD drawing.

3.2. Working Procedure

1. The plan layout of the Humayun Palace (Chepauk Palace) was obtainedas a CAD file.
2. A centerline drawing of the ground floor layout was drawn and this CAD file was exported to the ETABS software.
3. The material properties for masonry walls, wooden beams, woodenrafters, tile slab and brick slab were inputted.
4. The section properties were defined for the masonry walls, woodenbeams, wooden rafters, tile slab and brick slab.
5. Walls are designed with provisions for necessary arch and door openings.
6. Openings for arches and doors were made.
7. Main wooden beams of 400x800mm cross-section were drawn.
8. Small wooden rafters of size 76 x 127mm were used as secondary beams.
9. 300 mm thick slabs were used. They were assumed to be made ofbrick masonry.
10. All the catch basins in the structure were filled with walls.
11. This model was replicated for the first floor, with necessary modificationsmade with respect to the original plan layout.
12. Relevant loads were applied. The load combinations are as follows –

1.	Dead Load	Live load	
2.	Dead Load	Live Load	Seismic Load X direction
3.	Dead Load	Live Load	Seismic Load -Y direction
4.	Dead Load	Live Load	Seismic Load-

5.	Dead Load	Live Load	Seismic Load-
6.	Dead Load	Seismic Load X direction	
7.	Dead Load	Seismic Load -X direction	
8.	Dead Load	Seismic Load +Y direction	
9.	Dead Load	Seismic Load -Y direction	

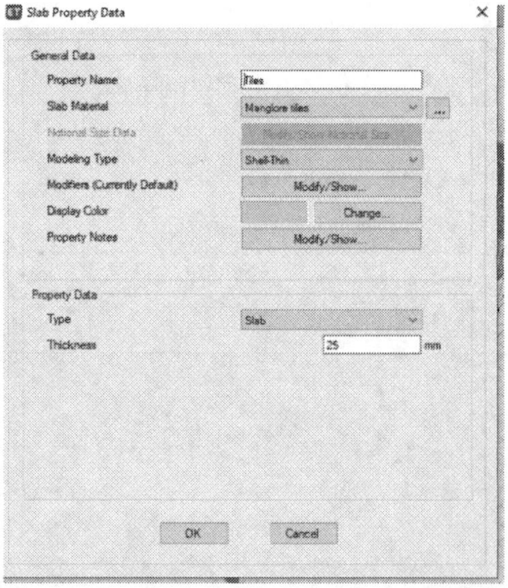

Fig.4.a Material and slab properties adopted during

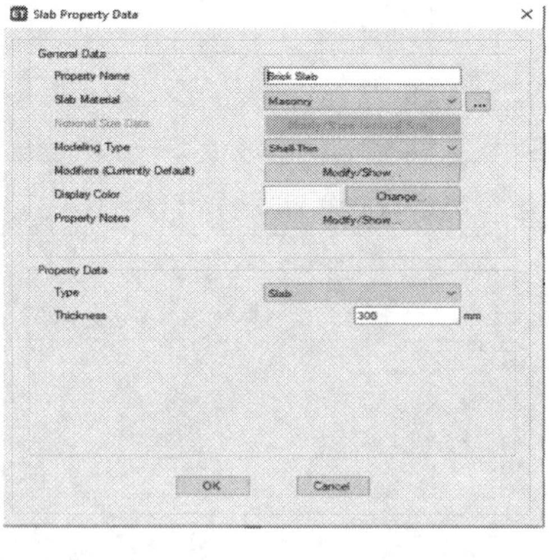

Fig.4.b. Material and slab properties adopted during

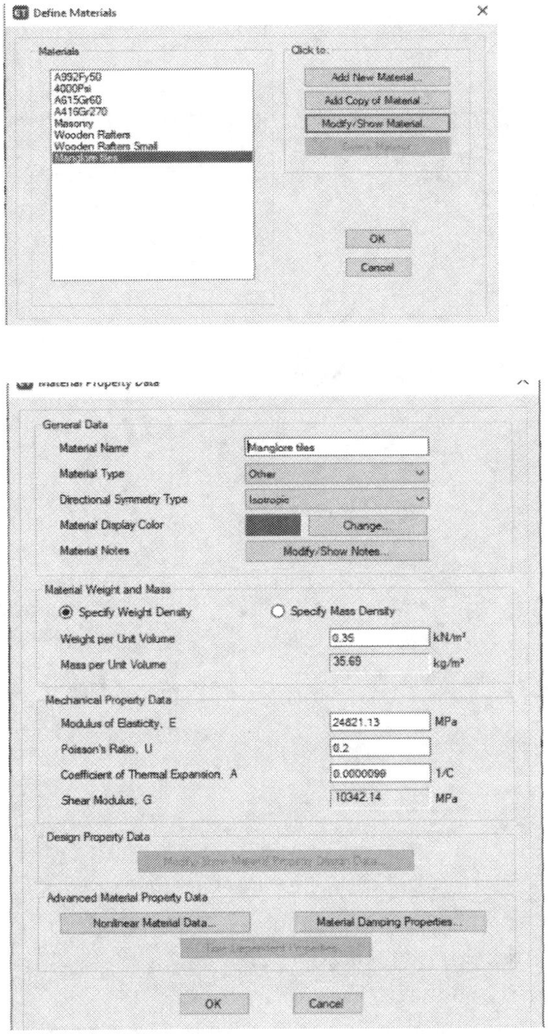

Fig.4.c. Material and slab properties adopted during analysis.

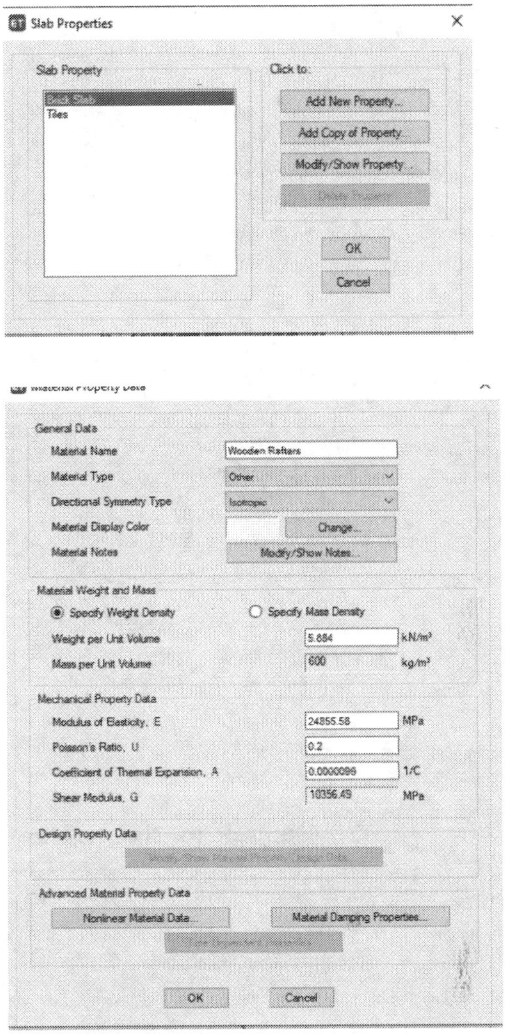

Fig.4.d. Material and slab properties adopted during analysis.

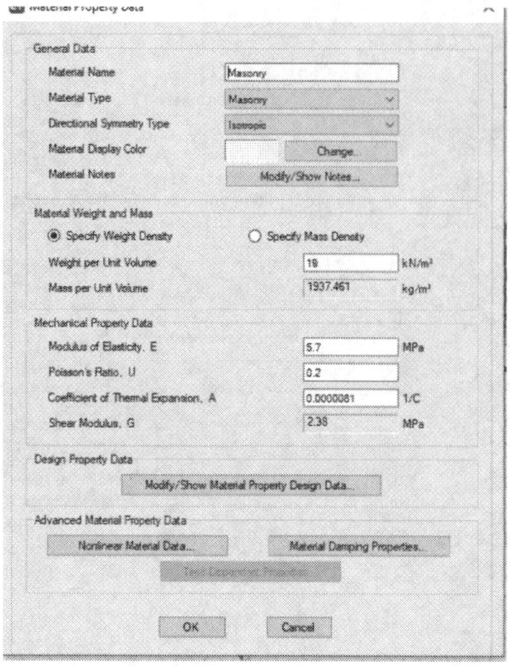

Fig.4.e. Material and slab properties adopted during analysis.

3.3. Seismic load considerations

- Seismic Loading was taken as per IS 1893:2016 Part 1. As, Chennai comes under Zone III as per the code, the following considerations were taken:
- Response Reduction Factor=1.5 for unreinforced masonry.
- Seismic Zone Factor = 0.1 for Zone III, shown in Fig.5.
- Importance Factor for important buildings such as monuments and government buildings
- Time period was calculated by the software.
- The seismic forces in the x and y directions are considered. Vertical seismic forces are ignored as the building is only two stories tall.

Table 3 Seismic Zone Factor *Z*
(*Clause* 6.4.2)

Seismic Zone Factor (1)	II (2)	III (3)	IV (4)	V (5)
Z	0.10	0.16	0.24	0.36

Fig.5 Seismic zone factor as per IS 1893

3.4. Combinations for Design of horizontal earthquake loadsAs Per IS Codes

1) 1.2 [DL+IL± (EL, ±0.3 EL,)] and 1.2 [DL+IL± (EL, ±0.3 EL,)].

2) 1.5 [DL± (EL, +0.3 EL,)] and 1.5 [DL (EL 0.3 EL,)]; and

3) 0.9 DL 1.5 (EL 0.3 EL) and 0.9 DL ± 1.5 (EL, +0.3 EL).

3.4.1.Combinations to be accounted to design of a three-dimensional structure according to IS 1983 2016

The seismic combination obtained the response due earthquake force (EL)is the maximum of the following three cases:

1) ± ELx±O.3 ELy±0.3ELz

2) ±ELy±O.3 ELx±0.3 ELz

3) ± ELz± 0.3 ELx±0.3 ELy

1) 1.2 [DL+IL+(EL, ±0.3 EL, ±0.3 EL,)] and
1.2 [DL + IL ± (EL, ±0.3 EL, ±0.3 EL)].
2) 1.5 [DL (EL, ±0.3 EL, 0.3 EL,)] and
1.5 [DL± (EL, ±0.3 EL, +0.3 EL,)]; and
3) 0.9 DL 1.5 (EL, ±0.3 EL, ±0.3 EL,) and
0.9 DL+ 1.5 (EL, +0.3 EL ±0.3 EL).

where x and y are two orthogonal directions and z is vertical direction.

Where,

$EL= \sqrt{} (ELx)^2 + (ELy)^2 + (ELz)^2$

3.5. Assumptions

13. Minor masonry projections were neglected.
14. Arches were taken as rectangular shapes. This was done as ETABS cannot factor in arch shapes and replacing it with a rectangular shapewouldn't cause much of a difference. Irregular shapes were taken as definitive shapes, such as squares or rectangles etc. This is illustratedin Fig.3. Fig 6(b) is the actual plan layout of the ground floor and Fig 6(a) is the center line diagram for the ground floor, which we took foranalysis purposes.
15. We considered the domes as concentrated loads.
16. We took the live load to be 3 kN/m.
17. We have considered the dome as a concentrated load and thecalculations are given below:

Diameter of the dome = 4.7m Thickness of the dome =100mm
Unit weight of brick is taken as 19Kn/m3 (as the dome is constructedusing brick masonry)

As we know, Surface area of hemisphere is

$$\frac{2}{3} \times \pi \times (r)^2 \, x \, (thickness\ of\ dome)$$

Load = unit weight of brick x is

$$\frac{2}{3} \times \pi \times \left(\frac{4.7}{2}\right)^2 x \ 0.1$$

By the equation, we got the concentrated load as 21.9Kn ~ $22kN$

Which we applied it as a gravity concentrated load to the structure.

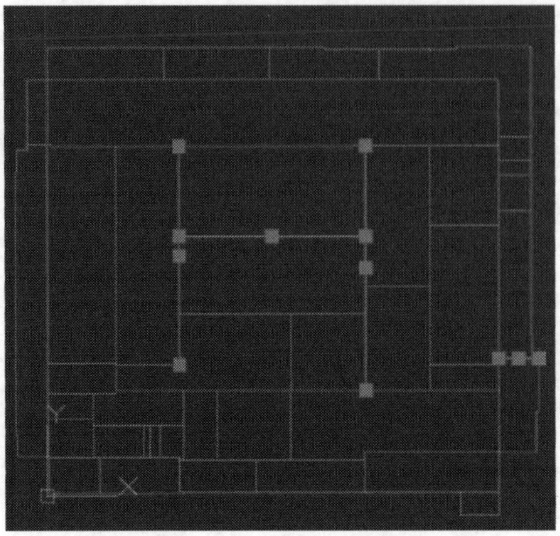

Fig.6.a. Centre line diagram of the ground floor used for analysis.

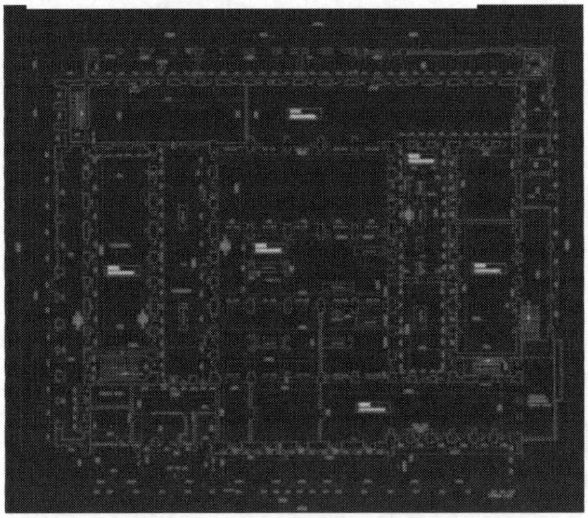

Fig.6.b. Actual plan layout of ground floor.

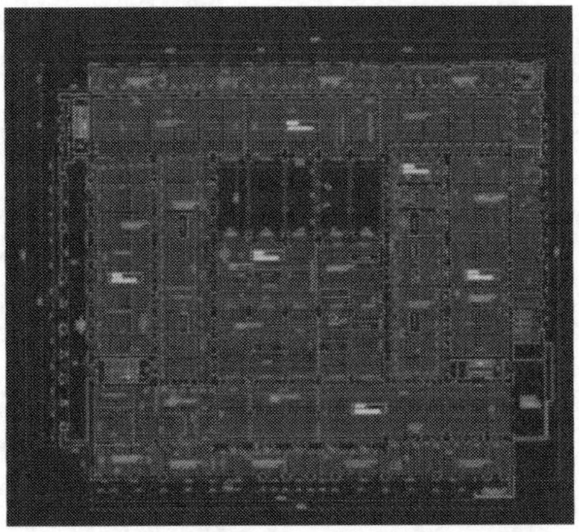

Fig.6.c. Ground floor top view CAD layout

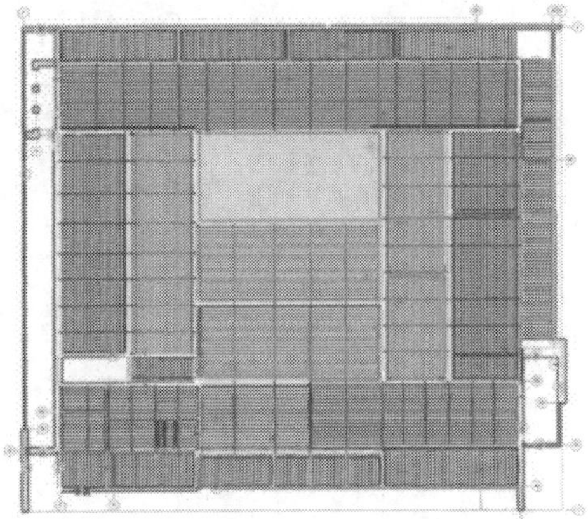

Fig.6.d. Modelled ground floor top view in ETABS

Fig. 6. Relative study of both the ground floor roof drawing.

3.6. Analysis

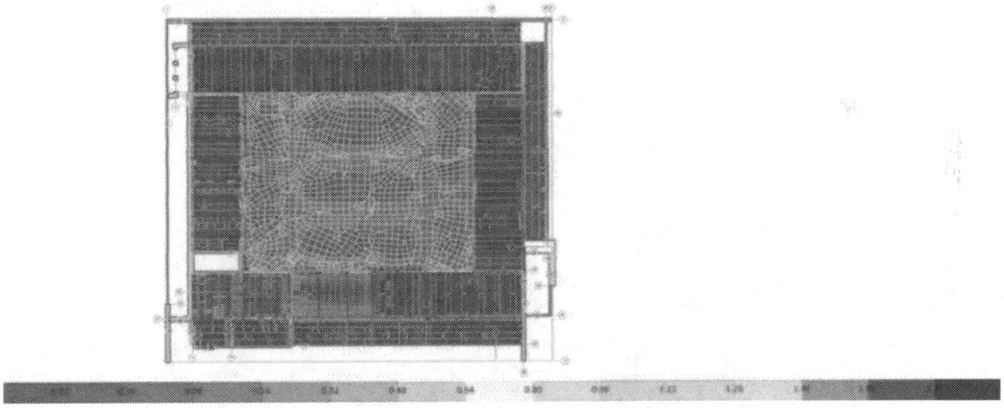

Fig.7.a. represents the maximum shell stress (kN/m2) distribution diagram in the plan view.

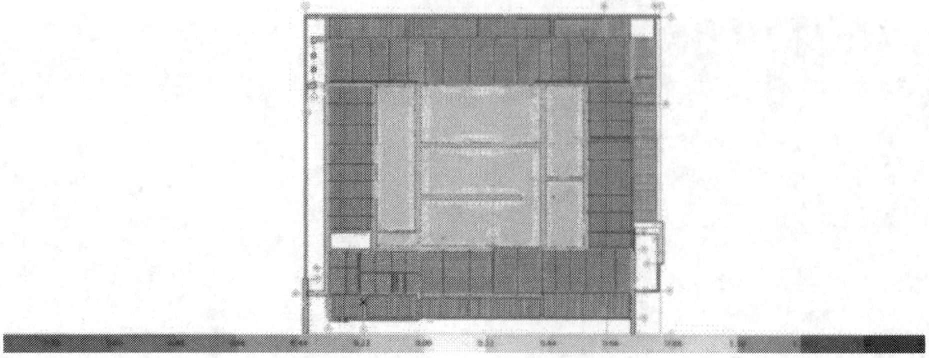

Fig .7.b. represents the average stress (kN/m²) distribution diagram inthe plan view.

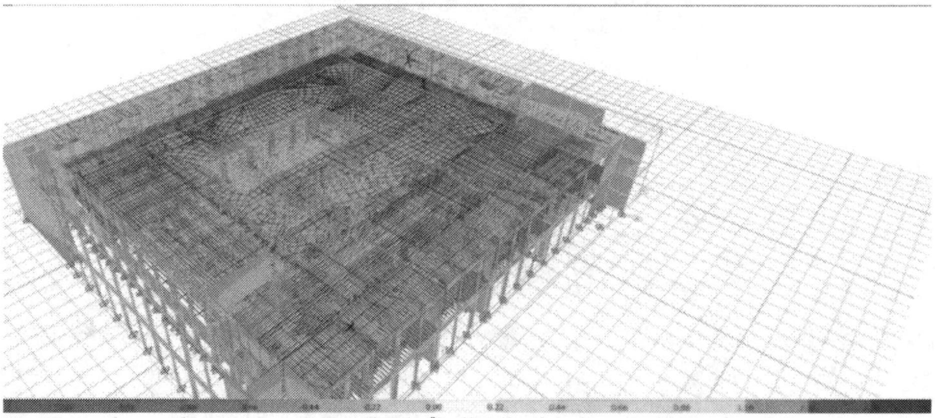

Fig.7.c. shows the average stress (kN/m²) distribution in a 3-dimensional format.

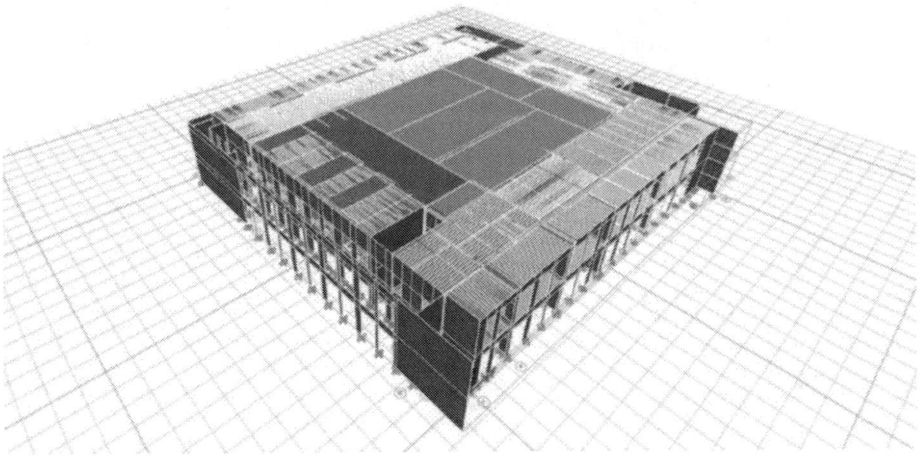

Fig. 7. d. shows the extruded 3-dimensional model and the 3-dimensional view if the model respectively.

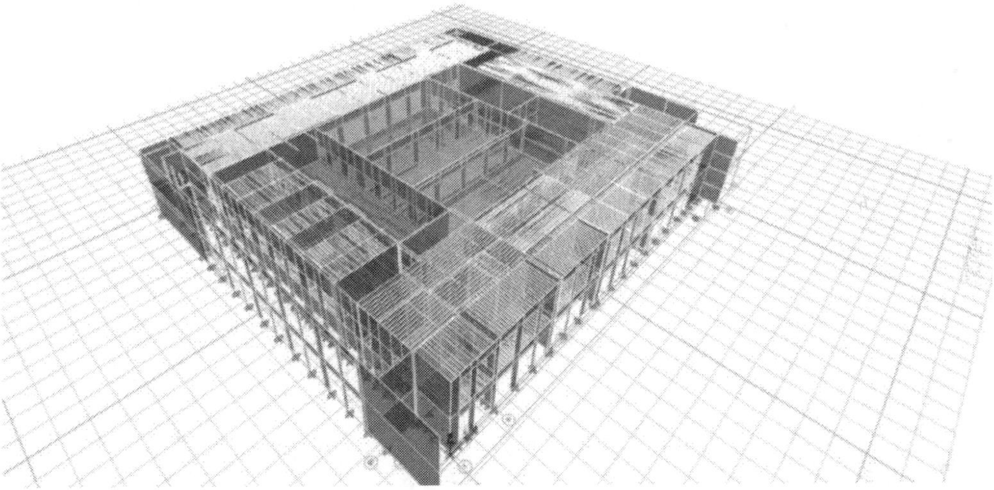

Fig. 7. e. show the extruded 3-dimensional model and the 3-dimensional view of the model respectively.

3.6.1. Important discussions from analysis

As from the analysis we could understand that the above shell stress/ forces contours obtained from its deformed shape under extreme envelope seismic design combinations. All the stresses are within the permissible limit which is ranging between 0.00 – 0.22 kN/m^2. So, the structure does not require any retrofits or seismic control systems for the building. The said structure was found to be structurally over designed and can withstand the present relevant seismic IS codes.

As mentioned already, arch openings are considered regular rectangular openings, which is why there aren't any failure zones observed there. If so, any failure zones were visible on the arch openings, and necessary seismic control systems can be provided depending on Indian needs,

The generally used seismic control systems are namely:

1. Viscous Dampers
2. FRP (Fibre Reinforced Polymer)
3. Steel Dampers
4. Bracing systems
5. Shape memory Alloys
6. Hybrid control Systems

Out of which hybrid control systems are the most commonly used seismic control methods, they achieve the performance of the building by control thebase displacement of the building.

4. Results and Discussions

The given heritage structure was modelled with the ETABS software using staticanalysis modelling. It shows that the shells have undergone minimal stress and strain. This may be attributed to the fact that proper foresighted planning was done. However, a few possible reasons may be deduced for the said results, thembeing: -

- Most modern commercial structures are constructed with 110 to 115 mm thickness slabs, but in this structure, the predominant thickness adopted was roughly 3 times more, i.e., 300mm.
- Usual walls constructed are 230 mm thick, but in this structure most wallsare of 400-1000 mm thickness.
- This structure also follows a different madras terrace slab system. Usually, rafters are placed in a certain centre-to-centre distance over which tiles are placed. But in this case, a main beam of comparatively bigger size is placed,and secondary beams of relatively smaller sizes are placed perpendicular to the same.
- Most importantly we have considered the arch openings are regular rectangular openings as there were some limitations working on E-Tabs if arch openings are incorporated, we might have identified some failure zones for the same and suitable retrofitting for the buildings can beprovided.

Following all these reasons, it is evident that the structure has been overdesigned, which capacitates it to have added stability and shows promising results. The Public Works Department has allocated nearly 96 crores for the aesthetic restoration of the structure, which was damaged during the fire of 2014.

5. Conclusions

Through this investigation, the value of the structure was brought out. It is evident that the E-TABS software and the static analysis method, though having a few restrictions, are brilliant means of simulations that can be used for further studies in monitoring or checking the health of the structure. The structure, through the ravages of time and deterioration, has stayed strong and shows promising results through the analysis. This can be owed to the overdesigning and sustainable provisions incorporated by the constructors during the time of construction. However, because of the restrictions the software laid, such as not allowing the actual shapes of the arches and other irregular shapes, the necessity to replace them with definitive shapes was inevitable. This may constitute the reason forthe structure not showing failure. In retrospect, if these provisions were to be incorporated, the structure may be prone to failure. The solutions to solve the issueof failure have also been furnished. Such cultural heritage sites must be preservedwith utmost care. The recent fire accident compounded the need to rehabilitate this structure, which has drawn the attention of the Public Works Department to take this project up for the same cause.

REFERENCES

[1] A.S, Khan R.A, Gupta H., Seismic Performance of a Masonry Heritage Structure IJEAT, Volume-3, Issue-4 (2014)

[2] G. Miceikaitė and J. Parasonis, Analysis of The Load Carrying Capacity The Masonry Walls of The Palace of Biržai Castle, The 9th international conference "Modern building materials, structures and techniques" (2017)

[3] S.S.L. Pelà and P. Roca, Nonlinear Numerical Modeling of Complex Masonry Heritage Structures Considering History-Related Phenomena in Staged Construction Analysis and Material Uncertainty in Seismic Assessment, J.
Perform. Constr. Facil., **34(5)** (2020)

[4] D. Dizhur, J. Ingham, L. Moon, M. Griffith, A. Schultz, I. Senaldi, Magenes, J. Dickie, S. Lissel, J. Centeno, C. Ventura, J. Leite and P. Lourenco Performance of Masonry Buildings and Churches in the 22 February 2011 Christchurch Earthquake, Bulletin of the New Zealand Society for Earthquake Engineering, Vol. 44, No. 4 (2011)

[5] J.A. Ronald, A. Menon, A.M. Prasad, D. Menon and G. Magenes Modelling and Analysis of South Indian Temple Structures Under Earthquake Loading, Sådhanå 43:74 Indian Academy of Sciences (2018)

[6] S. Sorace, G. Terenzi Structural Assessment of a Modern Heritage Building, Eng.Struct. 49, 743–755 (2013)

[7] J. Park, P. Towashiraporn, J.I. Craig, B.J. Goodnod, Seismic Fragility Analysis of Low-Rise Unreinforced Masonry Structures, Eng. Struct. 31, 125 137 (2009)

Laboratory study on partial replacement of conventional construction materials for the construction industry in India

Makendran C [1*] *Mahalingam B*[2], *and Jagan Inti*[3]

[1]Assistant Professor, Department of Civil Engineering, National Institute of Technology, Tiruchirappalli – 620015. Tamil Nadu, India
[2]Asscociate Professor, Department of Civil Engineering, Rajiv Gandhi Salai (OMR) Kalavakkam – 603 110, Tamil Nadu, India
[3]UG Student, Department of Civil Engineering, GMR Institute of Technology, Rajam, Andra Pradesh, India

Abstract: The study focuses on the conservation of natural resources by replacing three vital construction materials: (i) Ordinary Portland Cement (OPC) with Marble Dust (MD) (ii) Fine Aggregate (FA) with tile in powdered form and (iii) aggregate with tile chips (TC) for the building construction works. In this present study, the above-mentioned materials were used in 10, 20, and 30% as a replacement. As per the current scenario, only limited research papers are available that show what happens when the components are used as individual substitutions for the materials. Due to the lack of supporting evidence for material substitutions in concrete construction, the philosophy of experimenting with the use of mixtures of substitute materials in concrete has emerged. Experiments were conducted with M25 grade concrete, and the best mixes, M24 and M32, are mixes that replace all three traditional components of concrete, achieving strengths of 31.11, 2.76, and 8.0 N/mm^2 in compression, 33.33, and 2.76, and 8.25 N/mm^2 in flexure, respectively.

1 Introduction

Concrete is used more than any other material in the construction industry. Aside from water, it is also the second-most used substance globally. Concrete is a mixture of Fine Aggregate (FA) and Coarse Aggregate (CA) that are joined by a fluid cement that eventually becomes rigid. Most construction projects use either Portland Cement Concrete (PCC) or another type of concrete made with hydraulic cement [1]. The aggregate is mixed with the dry cement and water to form a fluid mass in the production of PCC and other hydraulic cement concretes,

* Corresponding author: makendran@nitt.edu

© The Authors, published by EDP Sciences. This is an open access article distributed under the terms of the Creative Commons Attribution License 4.0 (https://creativecommons.org/licenses/by/4.0/).

which is then easily moulded into the appropriate form of any structures [2]. The chemical processes that take place during the mixing of cement, water, and other elements produce a solid structure that binds all of them together to produce a durable substance that has a wide variety of applications [3]. Additives such as superplasticizers and pozzolanas are typically used in the mixture to improve the physical properties of the finished product or the wet mix. Reinforced concrete is created by pouring most concrete with reinforcing components like rebar inserted to enhance tensile strength [4]. From 45-65% of commercial structures in nations that are developing have marble floors. Due to the rapid expansion of the building industry and other construction activities, ingredients for concrete are currently in short supply [5]. These could affect ecological parameters and have a variety of negative effects on the environment, so need to find substitutes for them that will increase the mechanical characteristics of concrete more than conventional concrete, be more affordable than conventional concrete, and contribute to a decrease in the amount of waste dumped in landfills. In this present study, the aim is to conduct a Laboratory study on the partial replacement of conventional construction materials for the construction industry in India.

Various materials and methods have been recommended by different studies and researchers for determining the amounts of concrete constituents, using waste, recycling, and achieving essential strength. The study found this out by using marble dust as a further cementitious component in the building material. The study's findings indicate that concrete's strength has greatly increased [6]. It is recommended by some studies that concrete's coarse and fine aggregate may be replaced by ceramic tiles and conducted a cement concrete test by replacement of conventional materials. Cement was replaced by ceramic waste powder [7,8]. The replacement ranges from 0% to 30% of the compressive strength of concrete. Past researchers have done research work on concrete materials and related replacements of materials, especially in the Indian context, which is very less in number. Not many studies have empirically analysed the influence of the replacement of conventional elements used in concrete. The main objective of the construction industry in the study is to determine mechanical strength parameters through partial replacements, to determine the various percentages of partial substitution of cement with MD, fine aggregate with Tile Powder (TP), and coarse aggregate with TC, as well as to compare conventional and replacement concrete.

2 Methodology

The study focuses on the conservation of natural resources by replacing three vital construction materials: (i) Ordinary Portland Cement (OPC) with Marble Dust (MD) (ii) Fine Aggregate (FA) with tile in powdered form and (iii) aggregate with tile chips for the building construction works. This study consists of five stages. The first stage was the collection of raw materials and their processing into usable standards. The second stage was an engineering properties test for the chosen materials. The third stage was the mix design for various amounts of cement, FA, and CA are to be partially replaced with MD, TP, and TC respectively. Fourth-stage specimens were prepared and cured, and fifth-stage laboratory tests were conducted, and the results were analysed. Finally, the fifth-stage study conclusion and recommendations were announced.

2.1 Material Collection and Preparation

2.1.1 Cement

Ordinary Portland cement (OPC) is an extremely widely used type of cement. OPC constitutes 80–90% of the total cement production. In the study we conducted, an OPC of grade 53 was utilised and several cement tests were carried out. They include tests for consistency test, initial and final setting time, specific gravity, and other things as per IS :10500-2012.

2.1.2 *Fine Aggregate*

Sand from a nearby riverbank that is free of debris is utilised as FA. To reduce the void ratio as much as possible, the grains of sand should be packed as tightly as possible. When mixing, a higher level of voids necessitates the use of more water. Sand that conforms to zone III of Indian Standards was employed in the current investigation.

2.1.3 *Coarse Aggregate*

The crushed aggregate with a size of 20 mm and 12.5 mm was evaluated in accordance with Indian Standards, and the results were found to be within acceptable bounds.

2.1.4 *Marble Dust (MD) and Ceramic Tiles (CT)*

Marble Dust (MD) is a substance made from crushed or ground marble fragments in the marble industry. It is conceivable to use MD as alternative material for cement because a few studies have been done on its use as a substitute for part of cement which shows different Chemical Composition with MD & CT in Table 1.

Table 1 Different Chemical Composition with MD & CT

Chemical Composition	Ceramic Tile (CT) in (%)	Marble Dust (MD) in (%)	Remarks
SiO_2	63.29	13.8	Acceptable limit.
Al_2O_3	18.29	2.50	Acceptable limit.
Fe_2O_3	4.32	1.90	Acceptable limit.
CaO	4.46	43.20	Acceptable limit.
K_2O	2.18	0.60	Acceptable limit.
Na_2O	0.75	0.90	Acceptable limit.
MgO	0.72	2.70	Acceptable limit.
Cl	0.005	0.03	Acceptable limit.
SO_3	0.10	0.07	Acceptable limit.

Loss of Ignition (L.O.I)	1.61	43.63	Acceptable limit.
P_2O_5	0.16	N/A	Acceptable limit.
Mn_2O_3	0.05	N/A	Acceptable limit.

Source: Geo-Test House, Baroda, Gujarat, India, 2020

The substitute ingredients needed to make concrete are readily available in industrial form or as completed commodities. These final products, like tiles, cannot be used to prepare concrete directly. These materials must be treated to meet useable criteria prior to use.
Marble is utilised in the building construction as square tiles or floor slabs. You can't utilise this kind of marble to substitute for of cement. This means that before using marble as a replacement for cement, it must first be processed in accordance with the necessary criteria. Unevenly formed marble bits were gathered from a nearby collapsed building. The marble chunks were then ground into fine powder using an abrasion machine. It takes around 300 revolutions to turn 10-15 kg of marble chunks into fine powder. Next, a 300-micron sieve—the same size as a cement sieve—is used to sift the marble dust.

	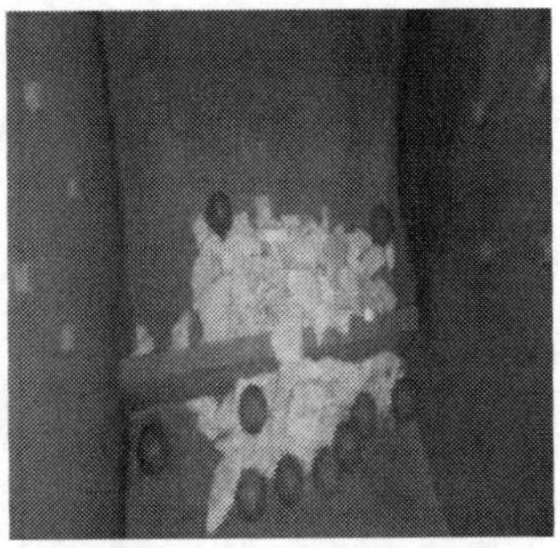	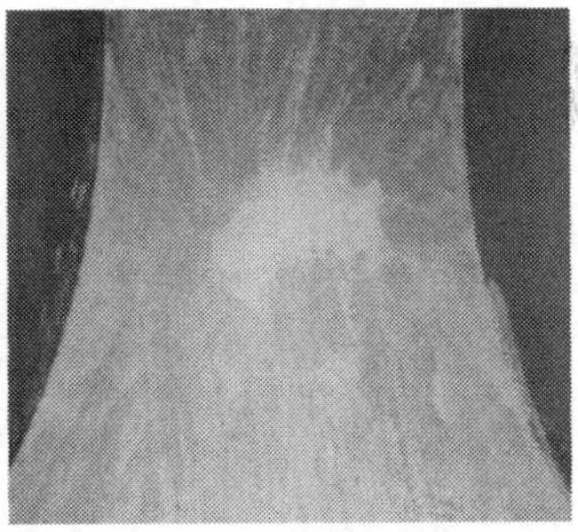
Fig 1(a) Source of marble pieces	**Fig 1(b) Loss angle testing machine set up with Marble pieces with steel balls (Before processing)**	**Fig 1(c) Marble dust obtained after processing**

*2.1.5 **Preparation of Tile Chips and Tiles Powder***

In nearby of Nidadavole, West Godavari District, Andhra Pradesh State, India. The broken tiles were gathered from a construction site. These bits were then further reduced in desired size using a hammer, sieved through an IS 20mm sieve, and utilised as a substitute for blue metal in concrete for percentages of 10, 20, and 30. TP was made by pounding tiles together, and the resulting sand-sized grains were sieved through an IS 2.36, whereas TC that made it through an IS 20mm sieve were collected. After these conditions were met, it was employed as a replacement for sand. Fig 2 shows the Tile chips and Tile powder.

	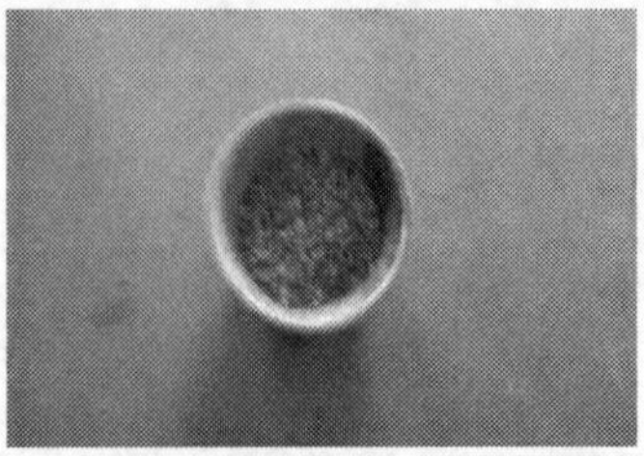
Fig 2(a) Tile chips sieved 20mm	**Fig 2(b) Tile powder passing 2.36 IS sieve**

2.1.6 *Specific Gravity (SG) and Water Absorption Test (WAT)*

Oven Dry Method (ODM) tests for the SG and WAT was performed for FA and CA in accordance with 3(2.4) of IS 2386-1963 part 1. The FA of SG and WAT were 2.48 and 3.67 respectively. The SG of CA for 12.5mm and 20mm was 2.55 and 2.84, respectively. While in WAT of CA for 12.5mm and 20mm was 0.98 and 0.603, respectively.

2.1.7 *Los Angeles Abrasion (LAA)*

To assess a coarse aggregate's LAA value, a test called the LAA is conducted. The test was carried out in accordance with IS: 2386 (Part IV) - 1963. LAA has a 61.7% abrasion.

2.1.8 *Test on Cement*

It was possible to determine the initial setting time of the cement by using the equipment made by Vicat. It was determined that the cement had a specific gravity of 3.14 and the initial setting time for the cement was 1 hour and 20 minutes.

2.1.9 *Water and Slump Cone Test*

A slump cone test was carried out to find out whether or not the concrete could be worked productively. According to Table 2 of the IS 10262: 2009 standard, it was found that 186 L of water are necessary in order to produce one cubic metre of concrete. In the case of 0.986 L, the workability was poor, and the slump value was 35 mm; on the other hand, in the case of 1.044 L, the workability was adequate, and the slump value was 100 mm.

3 MIX PROPORTIONING

Analyses were done using the design of the mix and IS 10262:2009 to determine the amount of concrete that would be necessary for constructing testing specimens using various proportions of the mix. Cubes, cylinders, and prisms are the cast specimens[13-18]. The amount of substitution components to be utilised in concrete was assessed based on the total weight of the traditional components utilized for concrete making. Ten, twenty, and thirty percent of the original elements are replaced. These materials have been utilised as substitutes for each of the three independent constituents as well as in combination. The table 2 and table

3 shows the M25 grade concrete mix ratio and percentage of alternative materials to their Corresponding Conventional Materials.

Table 2 M25 grade concrete Mix ratio

Materials	Unit	Replaced materials		Total Quantity
Cement	kg	90% of cement = 443.25	10% of Marble dust is 49.25	492.5
Fine Aggregate (FA)	kg	80% of FA = 462.32	20% of Tile Powder is 115.58	577.59
Coarse Aggregate (CA)	kg	70% of CA = 789.607	30% of Tile Chips is 338.403	1128.01
Water	Liters	226.59		
Mix Ratio	1.0: 1.17: 2.29			

Table 3 Percentage of alternative materials to their corresponding conventional materials

Mix	Replacement of Cement by Marble Dust (MD)	Replacement of Sand by Tile Powder (TP)	Replacement of Aggregate by Tile Chips (TC)
M1	MD0	TP0	TC0
M2	MD10	TP0	TC0
M5	MD0	TP0	TC10
M6	MD0	TP0	TC20
M7	MD0	TP0	TC30
M8	MD0	TP10	TC0
M9	MD0	TP20	TC0
M10	MD0	TP30	TC0
M14	MD10	TP20	TC10
M15	MD10	TP30	TC10
M16	MD10	TP10	TC20
M17	MD10	TP10	TC30
M24	MD10	TP20	TC20
M32	MD10	TP20	TC30

4 SAMPLE PREPARATION & TESTING

Prisms, cubes, and cylinders were the test specimens used for preparing the samples[13-18]. Nine cubes, a cylinder, and a prism were manufactured for each ratio. Three cubes were evaluated on the seventh day, and their average values were taken. The 14th day and 28th day findings were obtained using the same process. For 28 days, the cylinder and prism were put to direct testing. The cubes were 150 mm x 150 mm x 150 mm in size. The cylinder's nominal dimensions are 150 mm in diameter and 300 mm in height, while the prism's notional dimensions are 500 x 100 x 100 mm.

5 RESULTS AND DISCUSSIONS

5.1 COMPRESSIVE STRENGTH (CS)

Compressive strength is an essential parameter to consider when designing and evaluating structures. To measure CS, a universal testing machine is typically employed which is mentioned in IS: 456:2000. The cube test samples are 150 x 150 x 150 mm. The specimen is subjected to a prescribed rate of compression, typically between 0.5 and 2.0 MPa per second, until it breaks or fractures. The specimen's highest load capacity before failing was noted. The CS of a specimen may be measured by dividing its greatest weight by its cross-sectional area. At 28 days, the compressive strength was determined. In compressive strength, MIX1, MIX2, MIX5, MIX6, MIX7, MIX8, MIX9, MIX10, MIX14, MIX16, MIX17, MIX24, and MIX32 gave the best results 28.26, 31.55, 31.11, 34.44, 34.44, 36, 33.33, 28.89, 31.55, 31.11, 31.56, 31.11, and 33.33, respectively. Table 4 and Fig. 3 represent the entire compressive strength result.

5.2 SPLIT TENSILE STRENGTH (STS)

A material's ability to withstand tensile stress can be evaluated by measuring its tensile strength (TS), which is a type of mechanical property. It is defined as the TS of a cylindrical specimen that is subjected to a diametrical compressive load perpendicular to its longitudinal axis because tensile strength cannot be measured directly. So, the specimen has to be placed in a position such that the compressive load is applied along the entire length until failure occurs. The cylinder specimen is of size 300 mm in length and 200 mm in diameter. This is carried out within a compressive testing machine. The specimen is positioned horizontally between two steel plates that are secured to the pressure faces of the equipment being tested. A compressive load is applied to the specimen at a constant rate until it fractures into two halves. The load required to cause the specimen to fail is recorded, and the STS is measured. At 28 days, split tensile strength the was determined. In split tensile strength MIX1, MIX2, MIX5, MIX6, MIX7, MIX8, MIX9, MIX10, MIX14, MIX16, MIX17, MIX24, and MIX32 gave the slightly best results 1.41, 2.62, 2.62, 2.76, 2.48, 2.83, 2.26, 2.48, 2.55, 2.55, 2.41, 2.76, 2.76, respectively. Table 4 and Fig. 4 represent the entire split tensile strength result.

5.3 FLEXURAL STRENGTH (FS)

Flexural strength is a measure of how much bending a material can take before it breaks or breaks apart. It is worked out by putting weight on concrete prisms that are 50 cm by 10 cm by 10 cm and have a spread length that is at least three times the depth of the prism. FS is measured using established test procedures that conform to IS 516: 1959 and is represented as Modulus of Rupture (MR) in (N/mm^2). In flexure strength, MIX1, MIX2, MIX5, MIX6, MIX7, MIX8, MIX9, MIX10, MIX14, MIX16, MIX17, MIX24, and MIX32 gave the slightly best results, 6.25, 7.25, 7, 7.5, 6.75, 7.75, 7.25, 6.5, 8, 8, 7.25, 8, 8.25 respectively. Table 4 and Fig. 5 represent the entire flexural strength result.

Table 4 Compression, split tensile and flexural strength results in N/mm^2

Mix	Compressive strength (N/mm^2)	Split Tensile strength (N/mm^2)	Flexural strength (N/mm^2)
M1	28.26	1.41	6.25
M2	31.55	2.62	7.25
M5	31.11	2.62	7.00
M6	34.44	2.76	7.5
M7	34.44	2.48	6.75
M8	36.00	2.83	7.75
M9	33.33	2.26	7.25
M10	28.89	2.48	6.50
M14	31.55	2.55	8.00
M16	31.11	2.55	8.0
M17	31.56	2.41	7.25
M24	31.11	2.76	8.0
M32	33.33	2.76	8.25

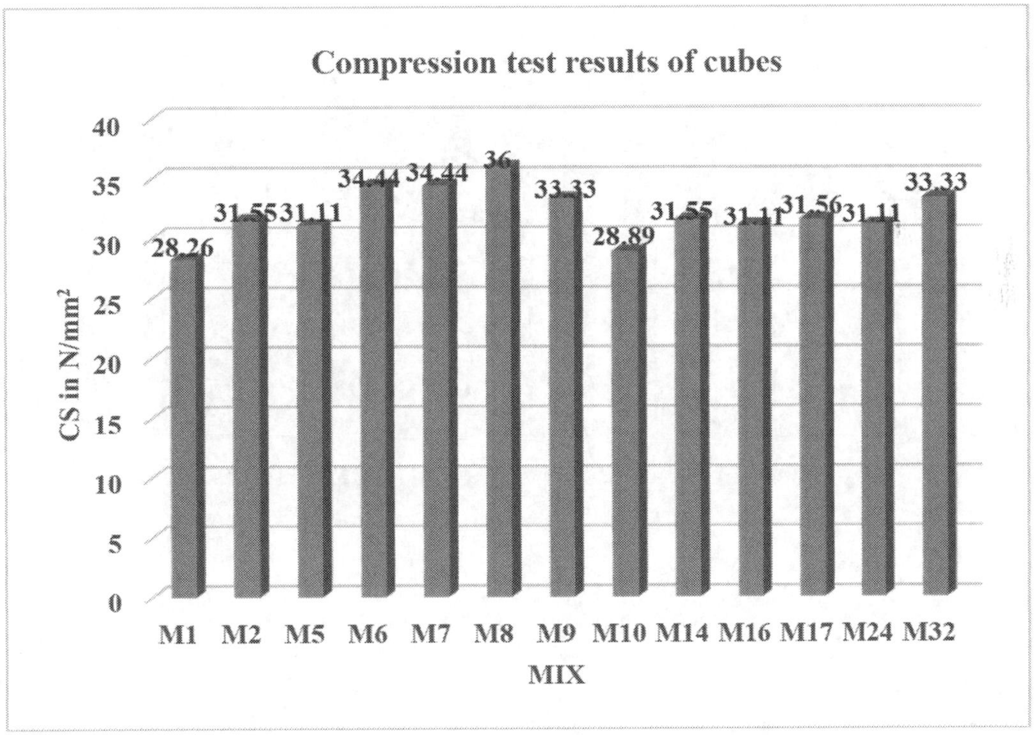

Fig. 3 Compression test results of cubes

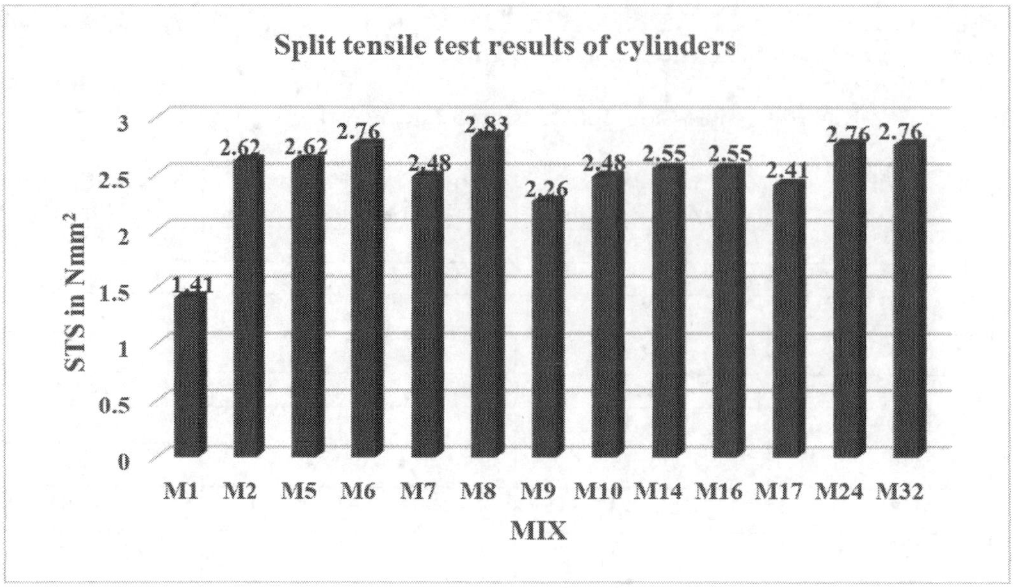

Fig 4 Split tensile test results of cylinders

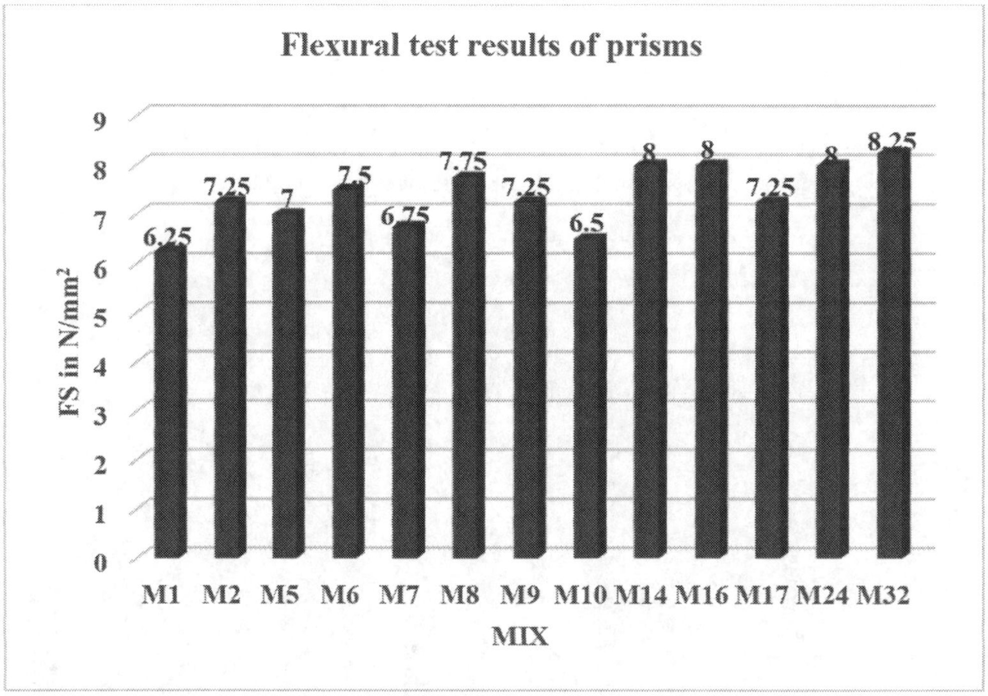

Fig 5 Flexural test results of prisms

6 CONCLUSIONS

The findings of the tests indicate that MD can effectively substitute up to 10% of the cement, TP can take the place up to 20% of the sand, and that TC can successfully take the place as much as 30% of the aggregate. An increase in the concrete's strength can be achieved by using the mixes MIX1, MIX2, MIX5, MIX6, MIX7, MIX8, MIX9, MIX10, MIX14, MIX16, MIX17, MIX24, and MIX32 also composites that include replacing all three standard components of concrete. These are referred to as composite mixes. This results in a concrete mix that is friendlier to the environment, as well as one that is more cost effective and stronger.

References

[1]. Muntadher J. Taher, E.H. Abed, Mahmood S. Hashim, Using ceramic waste tile powder as a sustainable and eco-friendly partial cement replacement in concrete production, Materials Today: Proceedings,2023, ISSN 2214-7853, https://doi.org/10.1016/j.matpr.2023.04.060.

[2]. Chao Du, Hongbo Tan, Shouwei Jian, Jie Wang, Tao Shi, Compressive strength and hydration process of sodium carbonate-activated superfine slag/marble powder binders, Journal of Building Engineering, Volume 43, 2021, 103121, ISSN 2352-7102, https://doi.org/10.1016/j.jobe.2021.103121.

[3]. K.P. Manikandan, P. Nanthakumar, M. Balachandar, D. Gowri Shankar, G. Vijayakumari, Partial replacement of aggregate with ceramic tile in concrete, Materials Today:Proceedings, 2023, ISSN, 2214-7853, https://doi.org/10.1016/j.matpr.2023.05.214.

[4]. Aviral Bagaria, Deepak Juneja, Experimental research on influence of marble powder, silica fume and polypropylene fiber on the porous concrete, Materials Today: Proceedings,2023, ISSN 2214-7853, https://doi.org/10.1016/j.matpr.2023.03.240.

[5]. Oumaima Bourzik, Khadija Baba, Nacer Akkouri, Abderrahman Nounah, Effect of waste marble powder on the properties of concrete, Materials Today: Proceedings, Volume 72, Part 7, 2023, Pages 3265-3269, ISSN 2214-7853, https://doi.org/10.1016/j.matpr.2022.07.184

[6]. Ghasan Fahim Huseien, Abdul Rahman Mohd Sam, Kwok Wei Shah, Jahangir Mirza, Effects of ceramic tile powder waste on properties of self-compacted alkali-activated concrete, Construction and Building Materials, Volume 236, 2020, 117574, ISSN 0950-0618, https://doi.org/10.1016/j.conbuildmat.2019.117574.

[7]. Parth R. Sondarva, J.R. Pitroda, Rajesh Gujar, Jaykumar Soni, An Experimental Investigation on the Strength Properties of Ceramic Tiles Waste Powder based Bacterial Concrete, Materials Today: Proceedings, Volume 62, Part 13, 2022, Pages 7062-7067, ISSN 2214-7853, https://doi.org/10.1016/j.matpr.2022.01.140.

[8]. Inti Jagan, Pongunuru Naga Sowjanya, Kanta Naga Rajesh, A review on alternatives to sand replacement and its effect on concrete properties, Materials Today: Proceedings, 2023, ISSN 2214-7853, https://doi.org/10.1016/j.matpr.2023.03.332.

[9]. Indian Standard Code of Practice for Plain and Reinforced Concrete (Fourth Revision) IS: 456-2000, Bureau of Indian Standards, July 2000, New Delhi.

[10]. Indian Standard Code of Practice for Method of Testing for Aggregate for Concrete, Particle Size and Shape (Eleventh Revision) IS: 2386(part I)-1963, Bureau of Indian Standards, October 1963, New Delhi.

[11]. Indian Standard Specifications for Portable Water IS: 10500-2012, Bureau of Indian Standards, May 2012, New Delhi.

[12]. Indian Standard Code of Practice for Splitting Tensile Strength of Concrete - Method of Test IS: 5816-1999, Bureau of Indian Standards, July 1999, New Delhi.

[13]. Chandrakasu, Makendran, Karunanidhi Suthandhiram, Shiferaw Garoma, Bekesha Merea, and Balaguru Sethuraman. 2022. "Laboratory Study on the Water-Soluble Polymer as a Self-Curing Compound for Cement Concrete Roads in Ethiopia" Technologies 10, no. 4: 80. https://doi.org/10.3390/technologies10040080.

[14]. Gujar Anantkumar Jotiram, Basanta Kumar Palai, Sumanta Bhattacharya, S Aravinth, G. Gnanakumar, Ram Subbiah, Makendran Chandrakasu, Investigating mechanical strength of a natural fibre polymer composite using SiO2 nano-filler, Materials Today: Proceedings, Vol 56, Part 3,2022, pp1522-1526, https://doi.org/10.1016/j.matpr.2022.01.176.

[15]. M.D. Mohan Gift, K.S. Raghuram, J.N.S. Suryanarayana Raju, D. Ravi, M. Venkata Rao, Ram Subbiah, Makendran Chandrakasu, Study on influence of nano-filler content on the performance of natural fibre reinforced epoxy composites, Materials Today: Proceedings, Vol 56, Part 3, 2022, pp.1562-1566, https://doi.org/10.1016/j.matpr.2022.01.304.

[16]. C. Makendran,Study on self-curing concrete using Polyethylene Glycol 600 for highway infrastructure project, Materials Today: Proceedings,Vol. 80, Part 2, 2023, pp.1677-1682. https://doi.org/10.1016/j.matpr.2023.02.316.

[17]. Makendran C.; Binaya Patnaik; Nilofer Hussaini; Jifara Chimdi Bikila; Ravindra Pathak; Khongdet Phasinam, Reliability analysis of cement manufacturing technique in computerized clinker processing method, AIP Conference Proceedings 2591, 030083 (2023) https://doi.org/10.1063/5.0119612.

[18]. T Udayabanu, NP Rajamane, C Makendran, R Gobinath, S Chandra Chary, Self-Curing Concrete Using Water-Soluble Polymerfor Developing Countries, IOP Conf. Ser.: Mater. Sci. Eng. 981 032088, 10.1088/1757-899X/981/3/032088

Studies on Estuary of Thamirabarani River, Tamilnadu, India

P.K. Suresh[1], N. Surendar[4], R. Sundaravadivelu[2], K. Dhanasekar[3]
[1] Consultant, Indian Institute of Technology, Madras, India-600036; Email: sureshpk2000@gmail.com
[2]Professor-Emertius, Indian Institute of Technology, Madras, India-600036 ; Email:rsun@iitm.ac.in
[3]Professor, Adi Shankara Institute of Engineering and Technology, Kalady,India-683574 ; Email: *dkollu@gmail.com*
[4]*Asst.Professor, Department of Civil Engineering, SSN College Kalavakkam, India – 603110;* Email: surendarn@*ssn.edu.in*

Abstract

The state of Tamilnadu is located along the south west of Indian peninsula. The state has nearly 40 river basin and most of them are seasonal. The river Thamirabarani is one of the perennial rivers covering 3 districts namely Tenkasi, Tirunelveli and Tuticorin. The river originates from the eastern side of the west coast and confluences with Bay of Bengal at a village called Punnakayal in Tuticorin district. The Thamirabarani river is known for its irrigation potential in the upper reaches. By the time it reaches the lower end, the flow is very low. The east coast is well known for its littoral drift which is dominantly towards the north direction. The village is famous for its fishing potential. The study area of Punnakayal is influenced by two monsoons namely south west and north east. Due to absence of high flow in the non-monsoon period, the estuaries are highly blocked by sand bar formation. Because of this sand bar formation, the fisherman find it difficult to go in to deep sea during low tide time. Hence studies were made and remedial measures were suggested and completed. The details of hydrologic analyses, hydraulic modelling, hydrodynamic studies, assessment of littoral drift remedial measures and post project performance are detailed in the paper.

Keywords: ***Drift, Estuaries, Hydraulic modelling, Thamirabarani river***

1. Introduction

The river Thamirabarani flows along Tirunelveli district of Tamilnadu and confluences with Bay of Bengal at Punnakayal (8.6349° N, 78.1151° E) is shown in Figure-1. The estuary was mostly blocked with sand bar formation. This is mainly due to long-shore littoral drift which is predominantly towards north along the coast and absence of perennial flows in the river. The place has good potential for fishing operations. Due to the formation of sand bar the fishermen are not getting even minimum draft to take their boats for fishing operations during low tide time.

1.1 Issues

The estuary was always blocked with sand bar formation. The place has good potential for fishing operations. Due to the formation of sand bar the fishermen are not getting even minimum draft to take their boats for fishing operations. Hence to manage the sand bar formations, studies are made and suitable remedial measures are carried out.

© The Authors, published by EDP Sciences. This is an open access article distributed under the terms of the Creative Commons Attribution License 4.0 (https://creativecommons.org/licenses/by/4.0/).

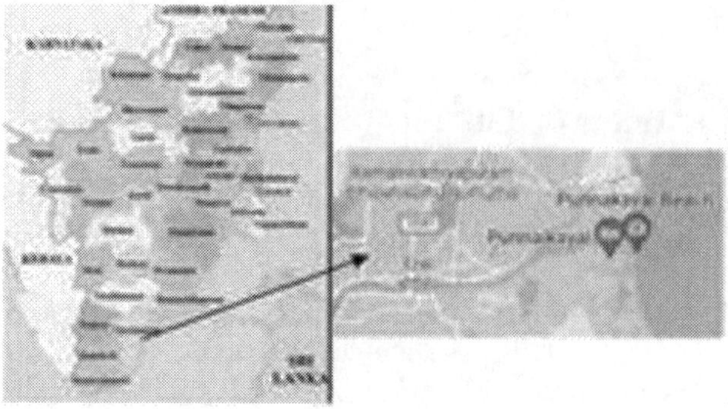

Figure-1 Location

2. Methodology

- Estimation of run off
- Tidal details
- Hydraulic modelling
- Wave hydrodynamic studies
- Estimation of littoral drift
- Design of remedial measures

2.1 Estimation of run off

The runoff estimation is made by delineating the catchment using GIS technique. The effects of catchment behavior on floods, prediction of catchment response to changes in the input conditions, modeling of river behavior, real-time flood forecasting, adjusting and evaluation of flood is assessed using geographic information systems (GIS). GIS can be defined as 'a computer system capable of capturing, storing, analyzing and displaying geographically referenced information that is, data identified according to location. The data for rainfall-runoff modeling ranges from a digital elevation model (DEM) to derive physical basin characteristics, soil characteristics. The advent of GIS, these data can be easily processed and analyzed to estimate hydrological modeling parameters from satellite images. The present location gets heavy rainfall only in north east monsoon. The Average annual rainfall 75cm with nearly 25 rainy days in a year. Usually, the location receives high rainfall of short duration. The Hydrologic Modeling System (HEC-HMS) is designed to simulate the precipitation-runoff processes of dendritic drainage basins. It is designed to be applicable in a wide range of geographic areas for solving the widest possible range of problems. This includes large river basin, water supply and flood hydrology, and small urban or natural watershed runoff. The maximum one-day rainfall received is 184mm. The discharge was estimated as 180 m^3/s adopting HEC-HMS (2016) tool.

2.2 Tide Details

Tidal levels play a major role in managing the ecosystem of an estuary. The tides observed at location are semi diurnal with ranging from 0.40 cm to 0.80cm is shown in Figure-2.

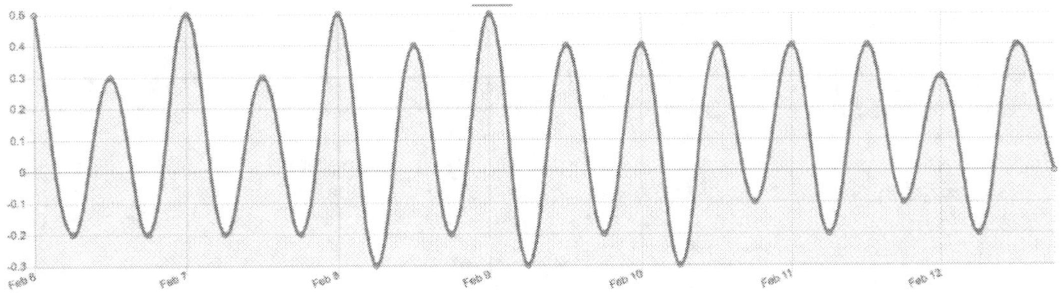

Figure-2 Tide Details

2.3 Hydraulic Modeling

The discharge carrying capacity of lower reach was estimated by hydraulic model studies. For the present study the U.S. Army Corps of Engineers' River Analysis System (HEC-RAS) software is used. This software is developed by the Hydrologic Engineering Center (HEC), which is a division of the Institute for Water Resources (IWR), U.S. Army Corps of Engineers. HEC-RAS is designed to perform one-dimensional hydraulic calculations for a full network of natural and constructed channels. The basic computational procedure is based on solving the one-dimensional energy equation using the Manning's equation in conjunction with the contraction and expansion coefficients using a procedure similar to the standard step method. The present analyses indicate no proper tidal exchange due to formation of sand bars during non-monsoon period from January to October. During heavy rains resulting more flow and thereby avoiding sand bar formations.

2.4 Wave Hydro-Dynamic Studies

A quantitative understanding of wave characteristics in the near shore is essential for the estimation of sediment transport and morphological changes along the coastal areas. Unfortunately measured or visually observed wave data is available only for locations of port. Hence, numerical models were resorted to for the simulation of wave climate. In the present study, two wave models of MIKE 21 (DHI, 2001), viz Offshore Spectral Wave (OSW) and Nearshore Spectral Wave (NSW) were used to derive the wave information in the near shore from the wind forcing. The physics and resolution of these models depend on the water depth for which the wave characteristics are derived (IITM 2010). The stretch of east coast of Punnakayal is found to be subjected to waves from south and south east directions can be seen in Figure 3.

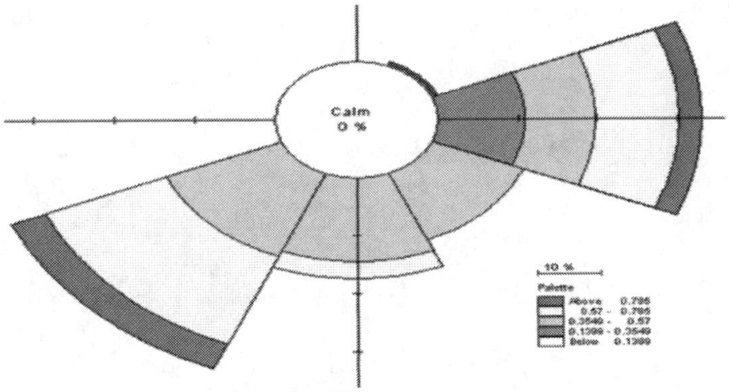

Figure 3 Wave rose

2.5 Estimation of littoral Drift

The study reach is always dominated by waves from south, south-west, south-East and east directions. The study indicates alongshore transport is towards north during which accretion of beach materials are observed on the south side of the shore protruding structure. The littoral drift estimates are made using Van Rijn (2001) formulation which considers both particle size and nearshore slope. The annual drift works out to be 0.060 to 0.080 million m^3 towards north along east coast. The data was validated with published estimations (Sundar & Suresh 2011). In the absence of continuous flow, the sediment transfer due to waves dominates and hence resulting in blocking of river inlets and formations of sand bar. This is the reason for blocking of present estuary.

2.6 Design of Remedial Measures

The estuary entrance sand bar formation can be avoided by constructing a pair of straight groynes on both sides of estuary(190m and 260m). The training walls should be taken sufficiently more than breaker depth to avoid accumulation of sand. The river bed should be dredged to a depth of (-) 2.50m to accommodate more discharge and for further propagation of tides.

2.6.1 *Project Completed*

The fisheries department of government of Tamilnadu has constructed two groynes at the estuary location and carried out dredging of bed to about (-) 2.50m in order to facilitate tidal propagation (Figure 4). The length of southern and northern groynes are 260m and 190m . A connecting bridge is also constructed upstream.

2.6.2 Post Project observations

Prior to the remedial measures there was no proper tidal propagation inside the river. This is because of the sand bar formation which prevents the propagation. In the post project scenario, the effect of semi diurnal phenomenon is seen. At present the tides propagate in such a way that two high water and two low water effects are seen.
The flow parameters were assessed based on the ground details provided by the clients. The slope of the river from near by Srivaikuntam anicut was estimated as about 1 in 2000. The study performed indicate that velocity was in the range of 1 to 1.5 m/s with average velocity of about 1.2m/s. After the remedial measures the velocity has increased. Near the bridge section it is around 5m/s which results in good flushing action during the flow.
Prior to the project the depth available was less than 1m. Only during the high tide the boats can perform the operation. But after the dredging about 3m draft is available in channelized area. During high tide it can go up to about 4m which can easily facilitate the boat movements

Figure-4 Training Groynes

3. Conclusions

The sand bar formation at estuary confluence point was controlled by constructing a pair of training walls in the form of groynes. The groynes are constructed after detailed studies and length was taken beyond breaker zone. At present there is good flushing action because channelizing by dredging and the high driving velocity near bridge section .The sedimentation is seen only on the south side of groyne located.

Acknowledgements

The authors acknowledge Er.Nambirajan, Assistant Engineer, Tamilnadu Water resources Department for his useful discussion pertaining to the study area.

References

Danish Hydraulic Institute DHI. (2001) User Manual and Reference Guide for LITPACK and MIKE21

Van Rijn, L.C (2001) "Longshore Sediment Transport" Delft Hydraulics

US Army corps of engineers Hydrologic Engineering Center HEC-RAS River Analysis System-Application guide March 2008.

P K Suresh (2010) Doctoral Thesis : "Nearshore sediment dynamics along the coast of Tamilnadu" I I T Madras

P K Suresh and V Sundar "Simulation of shoreline evolution and estimation of alongshore sediment transport" August 2011, in the ASCE Conference Proceedings. Solutions to Coastal Disasters August 2011

Hydrologic Modelling system HEC-HMS 2016, US Army corps of engineers Hydrologic Engineering Centre

AUTHOR INDEX